384.55 T267c

. .--mmuni-

ENTERED OCT 1 2003

TELECOMMUNICATIONS:
ISSUES IN FOCUS

COLUMBIA COLLEGE LIBRARY
600 S. MICHIGAN AVENUE
CHICAGO, IL 60605

TELECOMMUNICATIONS: ISSUES IN FOCUS

AGNES S. CORWALL (EDITOR)

Nova Science Publishers, Inc.
New York

Senior Editors: Susan Boriotti and Donna Dennis
Coordinating Editor: Tatiana Shohov
Office Manager: Annette Hellinger
Graphics: Wanda Serrano
Editorial Production: Jennifer Vogt, Ronald Doda, Matthew Kozlowski,
                          Jonathan Rose and Maya Columbus
Circulation: Ave Maria Gonzalez, Vera Popovich, Luis Aviles, Melissa Diaz,
                  Nicolas Miro and Jeannie Pappas
Communications and Acquisitions: Serge P. Shohov
Marketing: Cathy DeGregory

Library of Congress Cataloging-in-Publication Data
Available Upon Request

ISBN: 1-59033-356-X.

Copyright © 2002 by Nova Science Publishers, Inc.
              400 Oser Ave, Suite 1600
              Hauppauge, New York 11788-3619
              Tele. 631-231-7269   Fax 631-231-8175
              e-mail: Novascience@earthlink.net
              Web Site: http://www.novapublishers.com

All rights reserved. No part of this book may be reproduced, stored in a retrieval system or transmitted in any form or by any means: electronic, electrostatic, magnetic, tape, mechanical photocopying, recording or otherwise without permission from the publishers.

The publisher has taken reasonable care in the preparation of this book, but makes no expressed or implied warranty of any kind and assumes no responsibility for any errors or omissions. No liability is assumed for incidental or consequential damages in connection with or arising out of information contained in this book.

This publication is designed to provide accurate and authoritative information with regard to the subject matter covered herein. It is sold with the clear understanding that the publisher is not engaged in rendering legal or any other professional services. If legal or any other expert assistance is required, the services of a competent person should be sought. FROM A DECLARATION OF PARTICIPANTS JOINTLY ADOPTED BY A COMMITTEE OF THE AMERICAN BAR ASSOCIATION AND A COMMITTEE OF PUBLISHERS.

Printed in the United States of America

CONTENTS

PREFACE

Telecommunications is a field changing faster than our understanding of the words used to describe it: broadband, wireless, third generation and beyond, v-chips, access fees, encryption. This book brings together important data that details these and other major developments in the field.

384.55 T267c

Telecommunications

Chapter 1

RADIOFREQUENCY SPECTRUM MANAGEMENT: BACKGROUND, STATUS, AND CURRENT ISSUES

Lennard G. Kruger

INTRODUCTION

The radiofrequency spectrum, a limited and valuable resource, is used for all forms of wireless communication, including cellular telephony, radio, and television broadcast, telephone radio relay, aeronautical and maritime radio navigation, and satellite command, control, and communications. The radiofrequency spectrum (or simply, the "spectrum") is used to support a wide variety of applications in commerce, federal, state, and local government, and interpersonal communications. Because the spectrum cannot support all of these uses simultaneously to an unlimited extent, its use must be managed to prevent signal interference. The growth of telecommunications and information technologies and services has led to an ever increasing demand for the use of spectrum among competing businesses, government agencies, and other groups. As a result, the spectrum, which is regulated by the federal government, has become increasingly valuable. The need for managing the spectrum, including its allocation, has received growing attention by Congress in recent years.

SPECTRUM TECHNOLOGY BASICS

Electromagnetic radiation is the propagation of energy that travels through space in the form of waves. The most familiar form is light, called the visible spectrum. The **radiofrequency** spectrum is the portion of electromagnetic spectrum that carries radio waves. Figure 1 shows the radio spectrum as part of the measured electromagnetic spectrum. **Wavelength** is the distance a wave takes to complete one cycle. **Frequency** is the number of waves traveling by a given point per unit of time, in cycles per second, or **hertz** (Hz).[1] The

---

[1] Radiofrequency is usually measured in kilohertz (kHz), which is thousands of hertz, megahertz (MHz) which is millions of hertz, and gigahertz (GHz) which is billions of hertz.

relationship between frequency (f) and wavelength (λ) is depicted in Figure 2. **Bandwidth** is a measure of how fast data is transmitted or received whether through wires, air or space. Signals are transmitted over a range of frequencies which determines the bandwidth of the signal. Thus a system that operates on frequencies between 150 and 200 MHZ has a bandwidth of 50 MHZ.[2] In general, the greater the bandwidth, the more information that can be transmitted.

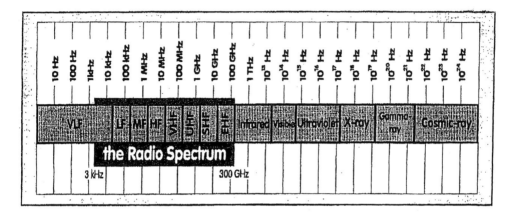

Figure 1. The Electromagnetic Spectrum

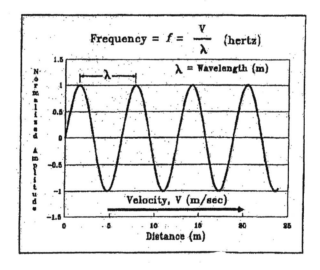

Figure 2. Frequency vs. Wavelength

An important distinction in spectrum technology is the difference between narrowband and broadband. **Narrowband** signals have a smaller bandwidth (on the order of kHz) and are used for limited services such as paging and low-speed data transmission. **Broadband** signals have a large bandwidth (on the order of MHZ) and can support many advanced telecommunications services such as high-speed data and video transmission. The precise

---

[2] Bandwidth is also measured in bits per second (bps) instead of cycles per second, especially in digital systems.

dividing line between broadband and narrowband is not always clear, and changes as technology evolves.

Two other important terms are analog signals and digital signals, depicted in Figure 3. In **analog** signal transmissions, information (sound, video, or data) travels in a continuous wave whose strength and frequency vary directly with a changing physical quantity at the source (i.e., the signal is directly analogous to the source). In **digital** signals, information is converted to ones and zeros that are formatted and sent as electrical impulses. Advantages of using digital signals include greater accuracy, reduction in noise (unwanted signals) and a greater capacity for sending information. Analog signals have the advantage of greater fidelity to the source, although that advantage can be made very small by increasing the rate at which signals are digitized. Digital signals are acknowledged to be superior to analog signals for the majority of applications.[3]

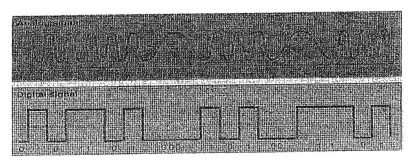

Figure 3. Schematic Comparing Analog vs. Digital Signals

Electromagnetic waves have many characteristics that govern how spectrum can be used in telecommunications systems. For example, antennas are used for transmitting and receiving signals, and can be designed to transmit in all directions or can be directed toward specific receivers. Receiving antennas are typically aligned with the transmitting antenna to maximize signal reception, but unintended signals can still interfere with the reception of the information sent. To avoid signal **interference** from stray signals, more than one radio signal usually cannot be transmitted in the same frequency range, at the same time, in the same area. Another characteristic is that the spectrum, unlike other natural resources, is not destroyed by use. As soon as one user stops transmitting signals over a portion of the spectrum, another can immediately re-use it. The spectrum is scarce, however, because at any given time and place, one use of a frequency precludes its use for any other purpose.

USES OF THE RADIO SPECTRUM

Spectrum is used to provide a variety of wireless communications services, which are categorized as fixed or mobile voice/data services or broadcast services. Demand for all

---

[3] For further discussion see CRS Report 96-401 SPR, Telecommunications Signal Transmission: Analog vs. Digital, May 7, 1996.

wireless services has grown rapidly in recent years in both terrestrial and satellite applications. Federal agencies use spectrum for various purposes, including military and national security needs, weather radio services, radars and communication systems to control commercial and private air and maritime traffic, weather satellite systems, flood warning and water control systems, and time signals. Examples include communications between defense platforms (e.g., aircraft and ships) and military bases, Voice of America broadcasts of politically oriented information in foreign countries, and data transmissions by the Department of Energy to monitor electrical power grids. Uses of the spectrum by state and local governments, and commercial entities are also varied and pervasive.

Much of the radiofrequency spectrum is shared among two or more wireless services. In these cases, one service may be designated as primary and the other services using the same frequencies as secondary, if the secondary services are prohibited from causing harmful radio interference to the operations of the primary service. For example, bands assigned to federal radar operations are primary, while at the same time, amateur radio spectrum is assigned secondary status within those same bands. Because these two systems do not interfere with each other, both may be designated as co-primary if they are given an equal degree of protection from harmful interference. Harmful interference is defined as any radio signal that degrades, obstructs, or interrupts the service provided by another operation.[4]

Commercial Voice and Data Transmission Services

- *Cellular telephone* systems consist of an array of terrestrial base stations (each covering a geographic area called a *cell*) that transmit and receive signals to and from mobile or fixed wireless telephones to provide two-way voice and data communications over a geographic region. Each cellular system is comprised of a cluster of cells of varying sizes according to the number of users in the area and the local terrain. All of the cells are connected to a mobile telephone switching office that manages all communications traffic in the cellular network and connects to public switched telephone network (the "wireline" network).

- *Paging* is a low-cost one-way message-sending system that uses base stations similar to cellular telephone systems. An enhanced paging system, called *messaging*, has a limited two-way data transmission capability.

- *Personal communications service (PCS)* is a wireless telephone services similar to cellular technology but using higher frequencies (around 1900 MHZ, compared to 800 MHZ for cellular services) and digital signal transmission technology (many cellular services are converting to digital). From the users' perspective, however, there is little difference between PCS and cellular services. Narrowband PCS can provide two-way messaging for interactive low-speed data applications (such as e-mail) but generally not voice, while broadband PCS provides a wider range of services.

---

[4] Code of Federal Regulations Title 47, Part 2, General Regulations, Sec. 2.1 Terms and Definitions.

- *Interactive video and data services (IVDS, not called 2-8-219 MHZ service)* is a subscription service that allows viewers cable and broadcast television to interact with the transmitting point over short distances. Applications include ordering goods or services offered by television programming, viewer polling, remote meter reading, vending inventory control, and cable television signal theft deterrence. These services now may also provide Internet access.

- *Specialized mobile radio (SMR)* is a wireless service originally created for public safety and dispatch communications. Newer enhanced SMR systems connect to the public telephone network to compete against cellular and PCS.

- *Satellite systems* provide communications to very large regions using signal transmissions between satellites and ground facilities. Communications satellites are used for voice, data, and broadcast purposes for government and commercial operations. Geostationary satellites maintain a fixed position relative to a point on Earth, and provide communications services for users at fixed or mobile locations. Low and medium Earth orbiting satellite systems are also being developed to provide fixed or mobile communications services including paging, voice, fax, and interactive Internet services.

Additionally, the Federal Communications Commission (FCC) has made spectrum available for unlicensed data services used for low-power applications. Providing this free spectrum for unlicensed data services is known to stimulate entrepreneurial activity, and use of this spectrum is intensive. Devices allowed to operate without a license[5] include cordless telephones, hearing aids, citizens band radio, consumer digital devices, industrial, scientific, and medical equipment, family radio service, and other innovations. Unlicensed spectrum is also used for wireless computing, whereby portable laptop computers interact with mainframes or wireless local area networks (LANs). In 1998, the FCC made an additional 300 MHZ in the 5 GHz band available for Unlicensed National Information Infrastructure (NII) devices to facilitate wireless access to the Internet and stimulate the development of new devices. Recently, the FCC has modified its rules to permit a greater use of spread spectrum devices (i.e., devices that operate by spreading the communications signal over a wide range of frequencies, thereby reducing the possibility of creating interference with other wireless systems). The FCC has also made extremely high frequency spectrum (above 40 GHz) available for future unlicensed applications and has proposed making additional spectrum available above 40 GHz.

Commercial Broadcast Services

- *Radio,* the oldest broadcast service, includes the AM (535 to 1,605 kHz) and FM (88 to 108 MHZ) bands within which licenses are assigned at specific frequencies for terrestrial radio broadcast stations. The same frequencies can be used by multiple

---

[5] These are known as Part 15 devices because the rules for the operation of these devices are provided in the Code of Federal Regulations Title 47, Part 15.

stations if the stations have sufficient geographic separation. The radio industry is developing new digital audio broadcasting (DAB) technology using the same spectrum bands, and the FCC is developing rules for DAB services.

- *Broadcast television* includes over 1600 currently licensed full service TV stations occupying 402 MHZ in the VHF and UHF bands. Each TV station has a 6 MHZ license. Television broadcasters are now starting to provide new digital television services using the vacant portion of the same band of spectrum.

- *Multipoint distribution service (MDS),* also called wireless cable, is a television broadcast system using digital encrypted signal transmissions in the microwave band (2 to 3 GHz), and providing up to 100 TV channels. MDS includes both single channel and multi-channel MDS *(MMDS)* applications (a more commonly used term). MDS now also offers subscriptions to high speed Internet access and data transmission services. A related service call instructional television fixed service (ITFS) operates under similar technology, but offers more educational programming. In addition, the FCC recently created a new service called Multichannel Video Distribution and Data Service (MVDDS) which uses a technology and system similar to MDS and is permitted to operate on spectrum to be shared with non-geostationary satellite orbit fixed satellite service.[6]

- *Direct broadcast satellite (DBS)* is a high-powered satellite television delivery system in which consumers receive programming by small (18 inch) receiving antennas in the 12.2-12.7 GHz band. It is related to the older direct-to-home (DTH) satellite television services that use large receiving antennas and operate in several other frequency bands.

- *Digital Audio Radio Services (DARS)* is a new high-fidelity radio service offered by several companies to be delivered to automobiles and other locations by geostationary satellites.

- *Local multipoint distribution service (LMDS)*, also called cellular television, is a new video distribution service for urban areas. Using a cellular architecture, LMDS can also provide two-way telephony (to compete with cellular telephone services), teleconferencing, telemedicine, and data services.

MANAGEMENT OF THE RADIO SPECTRUM

Since the beginning of the 20th century, when radio broadcast signals were first transmitted, it was realized that the spectrum was a public resource. Governmental entities have assumed responsibility for managing it to avoid interference. Spectrum is managed internationally by the International Telecommunication Union (ITU, a specialized agency of

---

[6] FCC Makes Spectrum Available for New Fixed Satellite Service at Ku Band; Seeks Comment on Licensing New Fixed Service at 12 GHz. FCC News Release November 30, 2000.

the United Nations located in Geneva, Switzerland). The ITU maintains a Table of Frequency Allocations which identifies spectrum bands for about 40 categories of wireless services with a view to avoiding interference among services. The ITU sponsors biannual World Radio Conferences to update the Table in response to changes in needs and demand for spectrum.

Once the broad categories are established, each country must allocate spectrum for various services within its own borders in compliance with the ITU Table of Frequency Allocations. In the United States, the Communications Act of 1934 established the FCC as an agency independent from the executive brand, to manage all non-federal government spectrum (which includes commercial, state and local government uses), and preserved the President's authority to mange all spectrum used by the federal government. The President also manages frequency assignments to foreign embassies and regulates the characteristics and permissible uses of the government's radio equipment. The President delegates this authority to the Assistance Secretary of Commerce for Communications and Information who is also Administrator of the National Telecommunications and Information Administration (NTIA).

The Communications Act of 1934 directs the FCC to develop classifications for radio services, to allocate frequency bands to various services, and to authorize frequency use. The Act does not, however, mandate specific allocations of bands for federal or non-federal use, which is usually decided through agreements between NTIA and the FCC (although Congress has occasionally directed the transfer of specific spectrum bands from federal to commercial use). The Act authorizes the FCC to grant licenses for radio frequency bands, but provides few details other than requiring that FCC rulings be consistent with the "public interest, convenience, and necessity." The Act authorizes the FCC to regulate "so as to make available…a rapid, efficient, nationwide, and worldwide wire and radio communication service with adequate facilities at reasonable charges, for the purpose of the national defense, and for the purpose of promoting safety of life and property."

The primary FCC offices that develop and implement spectrum policy are the Mass Media Bureau (which regulates all U.S. television and radio stations), the Wireless Telecommunications Bureau (which manages all domestic commercial wireless services except those involving satellite communications), the International Bureau (handling international telecommunications and satellite policies), and the Office of Engineering and Technology (developing spectrum allocations and policy, experimental licensing, spectrum management and analysis, technical standards, and equipment authorization). The FCC develops rules for spectrum use and other telecommunications regulation through lengthy proceedings in accordance with the Administrative Procedures Act. The FCC's Enforcement Bureau monitors the airwaves to ensure that non-federal users are complying with FCC rules, orders and authorizations.

The NTIA offices that focus on spectrum policy include the Office of International Affairs which represents U.S. interests in international fora, and the Office of Spectrum Management which develops policies and procedures for domestic spectrum use by the federal government. This entails developing long-range plans and war and readiness plans for spectrum use, chairing the Interdepartment Radio Advisory Committee (IRAC),[7] and representing the United States government at International Telecommunications Union

---

[7] The IRAC is composed of representatives of 20 major federal agencies who develop policies for federal spectrum use.

Conferences such as the World Radio Conferences (although other federal agencies also participate).

NTIA assigns frequencies and approves the spectrum needs for all federal government systems to support their mandated missions.[8] NTIA strives to improve federal spectrum efficiency by requiring federal users to use commercial services where possible, promoting the use of new spectrum efficient technologies, developing spectrum management plans, and collecting spectrum management fees (pursuant to congressional mandate). Since most spectrum is shared between government and private sector uses, NTIA (in conjunction with the FCC) is working toward increasing private sector access to the shared program. As a provision of the Omnibus Reconciliation Act of 1993 (47 U.S.C. 927), NTIA has begun to reallocate 235 MHZ of spectrum from federal government use to the private sector (195 MHZ of that amount has been reallocated and the remaining spectrum is scheduled for auction by the FCC in 2002).

Spectrum Auctions

Because two or more signal transmissions over the same frequency in the same location at the same time could cause interference (a distortion of the signals), the FCC, over many years, has developed and refined a system of exclusive licenses for users of specific frequencies.[9] Traditionally, the FCC granted licenses using a process known as "comparative hearings" and later using lotteries. After years of debate over the idea of using competitive bidding (i.e., auctions) to assign spectrum licenses, the Omnibus Budget Reconciliation Act of 1993 (47 U.S.C. 927) added Section 309(j) to the Communications Act, authorizing the FCC to use auctions to award spectrum licenses for certain wireless communications services (later expanded by the Balanced Budget Act of 1997). The main category of services for which licenses may be auctioned are called commercial mobile radio services (CMRS) which include PCS, cellular, and most SMR and mobile satellite services. CMRS providers are regulated as common carriers (with some exceptions) to ensure regulatory parity among similar services that will compete against one another for subscribers.[10] The FCC has the authority to conduct auctions only when applications are mutually exclusive (i.e., two licensees in the same frequency band would be unable to operate without causing interference with each other) and services are primarily subscription-based.[11] The FCC does not have authority to conduct auctions for licenses that have already been issued.

---

[8] Major federal spectrum users include the Departments of Defense, Justice, Transportation, Interior, Agriculture, Commerce, Treasury, Energy, the National Aeronautics and Space Administration, and the Federal Emergency Management Agency.

[9] Technically, two signals will interfere with each other even if they are not at the same exact frequency, but are close in frequency. To avoid harmful interference, the frequencies must have frequencies that are sufficiently different, known as a "minimum separation."

[10] Other services, classified as Private Mobile Radio Services (PMRS), are prohibited from connecting to the public switched telephone network.

[11] Licenses are issued for the use of bands of spectrum. In general, a greater bandwidth can carry more information than a smaller bandwidth.

Auction Rules

The FCC initially developed rules for each auction separately (with some common elements), but after several years of trial and error, it developed a set of general auction rules and procedures. While there may be special requirements for specific auctions, the following rules generally apply. As a screening mechanism, all auctions require bidders to submit applications and up-front payments prior to the auction. Most auctions are conducted in simultaneous multiple-round bidding, in which the FCC accepts bids on a large set of related licenses simultaneously using electronic communications. Bidders can bid in consecutive rounds on any license offered until all bidding has stopped on all licenses. Even though licenses must be renewed periodically, it is generally understood that license winners will be able to keep the license perpetually, as long as they comply with FCC rules.[12]

For some auctions, the FCC gives special bidding credits to smaller companies, called entrepreneurs, defined as having annual gross revenues of less than $125 million and total assets of less than $500 million. In the first year or so of auctions, the FCC originally also gave special provisions to women-owned, minority-owned, and rural telephone companies (called *designated entities*). After a 1995 Supreme Court decision determined that government affirmative action policies must pass a "strict scrutiny" test to demonstrate past discrimination, the FCC removed those other groups from its list of businesses qualifying for bidding credits.[13] Nevertheless, concerns have been raised that some of the small businesses participating in auctions actually represent larger companies that are excluded from the bidding credits.

Service Rules

The FCC also develops services rules for each new service for which a license will be used. Licenses are granted according to the amount of spectrum and the geographic area of coverage. The FCC's plan for the amount of spectrum per license, the number of licenses, and the conditions for use of the designated spectrum, known as the "band plan," is developed for each new wireless service. Licenses can cover small areas, large regions, or the entire nation. Terms used for coverage areas include basic trading areas (BTAs) which correspond roughly to metropolitan areas; major trading areas (MTAs), which are combinations of BTAs dividing the United States into 51 geographic regions of similar levels of commercial activity; and regions, which are combinations of MTAs. Metropolitan statistical areas (MSAs), rural service areas (RSAs), economic areas (EAs), and major economic areas (MEAs) developed by the Department of Commerce for economic forecasts are also used by the FCC to define areas of coverage for some spectrum auctions.

The FCC has also modified some wireless service rules to help new spectrum licensees maximize the value from their licenses. Changes include allowing licensees to partition licenses for greater efficiency, sharing regions among licensees, and expediting the relocation of incumbent microwave licensees from the spectrum purchased in the PCS auctions. The FCC maintains a website on its auction activities at [http://www.fcc.gov/wtb/

---

[12] The FCC provides additional information on auctions on its website at http://www.fcc.gov. wtb.auctions.htm.
[13] Adarand Constructors Inc., petitioner v. Federico Pena, Secretary of Transportation, et al. Docket No. 93-1841, decided June 1995.

auctions]. This site provides archived information on all of its completed auctions, details on its ongoing and future auctions, auction-related maps, charts, and service, and auction rules.

Complications with Some Auctions[14]

Despite their general success, the FCC's auctions have experienced several problems from which the FCC has learned and modified subsequent auctions.

WCS Auction

By 1997, the FCC had raised over $22 billion from auctions, and many observers in government and the private sector claimed auctions to be a success. That enthusiasm decreased somewhat, however, after the results of two auctions, held in April 1997, for wireless communications services (WCS) and digital audio radio service (DARS). The WCS auction was mandated by the FY1997 Omnibus Appropriations Act (P.L. 104-208 Title III), which directed the FCC to reallocate the use of 30 MHZ of spectrum (some of which had already been allocated for DARS) and to begin the auction for those licenses by April 15, 1997. To implement the Act, the FCC divided the spectrum remaining for DARS in half, surrounded by the newly allocated WCS spectrum. To prevent interference between the two new services, the FCC placed restrictions on the power that could be radiated by WCS. Although the FCC completed the DARS and WCS auctions in April 1997, meeting the congressionally mandated schedule, the revenue obtained was far lower than estimates had previously predicted. Reasons cited for the shortfall included the shortened timetable set by Congress for the FCC to complete the auction, and the technical constraints placed on the WCS spectrum which prevented interference with DARS, but reduced the usefulness of the WCS spectrum.

LMDS Auction and Satellite Spectrum Allocations

The auction for Local Multipoint Distribution Service (LMDS), a new television distribution service for urban areas that may also be used for two-way communications, was held in March 1998. LMDS uses much higher frequencies than existing commercial wireless services (in the 28 and 31 GHz bands). LMDS could compete with cable TV, broadcast TV, MMDS, satellite TV, mobile telephone data, or broadband Internet access services. Finding spectrum for LMDS was complicated due to the spectrum needs of new satellite services in the same bands. At the time when FCC was developing plans for LMDS spectrum, several companies developing new fixed satellite services (FSS) requested the same spectrum for sending signals to their satellites. FSS systems use geostationary and non-geostationary satellites to provide worldwide voice, video, and interactive data services to users at fixed locations. Mobile satellite services (MSS), which serve mobile users as well as fixed users, also wanted these frequencies to interconnect MSS systems to other communications networks, and the FCC had already granted licenses to several of these companies. The FCC

---

[14] See archived CRS Report 97-218, Radiofrequency Spectrum Management, updated April 23, 1998, for further analysis of these auctions and other spectrum management issues from a historical perspective.

decided not to auction the spectrum allocated for FSS or MSS, but divided the 28 GHz band among LMDS, geostationary FSS, non-geostationary FSS, and MSS, in the auction rules for LMDS licenses.[15] A frequency band around 18 GHz was designated for FSS downlink signals to share with several other services. The first LMDS auction raised $577 million; lower than expected by many in the private sector. Because some LMDS licenses received no bids, and other LMDS license winners defaulted on their payments, in 1999 the FCC conducted a re-auction of the remaining and reclaimed licenses, raising an additional $45 million. The reasons for the lower than expected proceeds are not clear, but could be due to a downturn in the market for spectrum at the time of the auction.

C-Block Auction

The auction of one of the blocks of spectrum allocated for PCS, known as the *C-block*, has presented some complex legal problems for the FCC. In the original C-block auction, also called the entrepreneur's auction, the FCC gave bidding credits to small businesses to help them compete with larger entities in the auction. Winning bidders only had to pay 10% down and the remainder could be paid over ten years at below-market interest rates. Although the initial auction, completed in May 1996, raised $10.2 billion, by mid-1997 many of the license winners (most notably NextWave Telecom, Inc.) defaulted and declared bankruptcy. The licenses were then seized by a court in bankruptcy litigation. In September 1997, the FCC offered a set of options for C-block licensees to restructure their debt (that offer was modified in March 1998). However, the licensees opted to maintain their bankrupt status, preventing the C-block spectrum from being re-auctioned. As a result of a series of decisions in 1999 and 2000 by the U.S. Court of Appeals, the FCC was ultimately able to cancel and re-auction the licenses. The auction for the defaulted licenses was completed January 26, 2001, and yielded $16.86 billion in revenue.

However, on June 22, 2001, the United States Court of Appeals for the District of Columbia found that the FCC did not have the legal right to take back NextWave's licenses for re-auction, and that 216 of the licenses (which garnered $15.85 billion in the auction) still belonged to NextWave rather than re-auction winners such as Verizon Wireless. Possible next steps include further litigation or a negotiated settlement between the FCC, NextWave, and auction winners.

To avoid future problems similar to those experienced in the C-block auction, in December 1997 the FCC adopted streamlined auction rules for all services to be auctioned in the future.[16] The rule changes were intended to ensure uniform procedures involving the application, payment, and certain concerns regarding designated entities (i.e., small businesses, women, minorities and rural telephone companies). For example, in many cases the FCC specifies a minimum opening bid prior to an auction, and provides more time prior to the auction for potential bidders to develop business plans, assess market conditions, and evaluate the availability of equipment. The FCC also recommended legislation, which was not enacted (see **Spectrum Auction Procedures** and **Issues for Congress**).

---

[15] FCC CC Docket 92-297 Fourth NPRM and First Report and Order, on Domestic Public Fixed Radio Services, released July 22, 1996, amended by an Order on Reconsideration, released May 16, 1997, and Second Order on Reconsideration, released September 22, 1997 to Establish Rules and Policies for LMDS and FSS.

[16] FCC 97-413, WT Docket 97-82, ET Docket 94-32, Third Report and Order and Second Further NPRM on Streamlining Auction Rules, released December 31, 1997.

800 MHZ SMR Auction

Another FCC auction that was criticized by some wireless service providers was the auction of SMR licenses in the 800 MHZ range, completed in December 1997. The FCC originally envisioned the 800 MHZ SMR licenses to be similar to those created in the 900 MHZ SMR auctions. The main difference, however, was that many more incumbent SMR licensees existed in the 800 MHZ band than were in the 900 MHZ band. The incumbents were not only concerned about potential interference, but also that they would never again be able to request additional spectrum from the FCC to expand their services. After much contention in a proceeding that lasted three years, the FCC adopted rules for the 800 MHZ licenses, despite the continued dissatisfaction of incumbent SMR licensees. The FCC required incumbents to relocate (against their wishes) to other frequencies after a mandatory negotiation period with new SMR licensees. The new licensees would have to pay for the relocation, but incumbents were forced to compete with the new SMR licensees for the incumbents' existing customers. A total of 524 licenses were sold in the auction, with one large SMR Company, Nextel, winning 90 percent of the new licenses. Many claimed that smaller SMR providers were not able to compete against Nextel in the auction. A similar set of issues surrounded the FCC's auctions for 220 MHZ licenses, paging services, and location monitoring services.

Incidents of Collusion

In the PCS auction of D, E, and F Block licenses, held in late 1996, some competing bidders were accused of using unusual bid amounts as a means of signaling their market intentions to each other during the auction. By early 1997, the Department of Justice began an investigation into bidding practices employed by participants of the PCS auctions. Based on this investigation, the FCC found specific parties liable for violating FCC auction anti-collusion rules that prohibit bidders from sharing their bidding strategies with competing bidders.[17] The FCC has since modified its bidding procedures so that all bids must be made in specific increments instead of any dollar amount to prevent collusion. It is still possible, however, for bidders to use other forms of collusion to keep prices low.

The Balanced Budget Act of 1997

The Balanced Budget Act of 1997 (47 U.S.C. 153) contained several spectrum management provisions. It amended Section 309(j) of the Communications Act to expand and broaden the FCC's auction authority and to modify other aspects of spectrum management. Whereas previous statutes gave the FCC the authority to conduct auctions, the Balanced Budget Act requires the FCC to use auctions to award mutually exclusive applications for most types of spectrum licenses. Exempted from auctions are licenses or construction permits for:

---

[17] FCC-98-42, Notice of Apparent Liability for Forfeiture for Facilities in the Broadband PCS in the D, E, and F Blocks. Adopted March 16, 1998.

(A) public safety radio services, including private internal radio services used by state and local governments and non-government entities and including emergency road services provided by not-for-profit organization, that –
   (i) are used to protect the safety of life, health, or property; and
   (ii) are not made commercially available to the public;

(B) digital television service given to existing terrestrial broadcast licensees to replace their analog television service licenses; or

(C) noncommercial educational broadcast stations and public broadcast stations.

Examples of services exempted from auctions include utilities, railroads, metropolitan transit systems, pipelines, private ambulances, volunteer fire departments, and not-for-profit emergency road services. This section also extends the FCC's auction authority to September 30, 2007. It directs the FCC to experiment with combinational bidding (i.e., allowing bidders to place single bids on groups of licenses simultaneously), and to establish minimum opening bids and reasonable reserve prices in future auctions unless the FCC determines that it is not in the public interest.

Furthermore, the Act directed the FCC to use auctions for mutually exclusive applications for new radio or television broadcast licenses received after June 30, 199. For applications filed prior to that date, bidding was limited to those who had already filed. Previously, the FCC granted all broadcast licenses through comparative hearing procedures. After a 1993 court case in which FCC criteria for selecting license winners were challenged, however, the FCC had stayed all ongoing comparative hearings pending resolution of the case. Following enactment of the Act, the FCC established auction procedures for all licenses for new commercial radio and television stations, as well as competing applications for new stations filed before July 1, 1997.[18] The FCC has since then conducted several auctions for broadcast licenses and broadcast station construction permits.

The Act directed the FCC to auction 120 MHZ of spectrum, most of which had already been transferred by NTIA from federal to non-federal use. As a result of this provision, the FCC began a proceeding on the allocation and auction of 45 MHZ between 1710-1755 MHZ 9pending), and must allocate by September 2002 another 55 MHZ located below 3 GHz for auction. It also directed NTIA to reallocate another 20 MHZ below 3 GHz (reduced to 12 MHZ be subsequent legislation – see **Allocation of Spectrum for Federal vs. Commercial Use**) for commercial uses. The Act also authorized private parties that win spectrum licenses encumbered by federal entities to reimburse the federal entities for the costs of relocation if the private parties want to expedite the spectrum transfer.

The Act required the FCC to conduct auctions for 78 MHZ of the analog television spectrum planned to be reclaimed from television broadcasters at the completion of the transition of digital television. That spectrum is to be auctioned in 2002 but not reclaimed from broadcasters until at least 2006. It then directs the FCC to grant extensions to stations in television markets where any one of the following three conditions exist: (1) if one or more of the television stations affiliated with the four national networks are not broadcasting a digital

---

[18] FCC 98-194, MM Docket No. 95-31, Implementation of Section 309(j) of the Communications Act – Competitive Bidding for Commercial Broadcast and Instructional Television Fixed Service Licenses, released August 18, 1998.

television signal, (2) if digital-to-analog converter technology is not generally available in the market of the licensee, or (3) if at least 15% of the television households in the market served by the station do not subscribe to a digital "multi-channel video programming distributor" (e.g., cable or satellite services) and do not have a digital television set or converter. To maximize the pool of potential bidders in auctions of the returned analog TV spectrum, the FCC may not disqualify bidders due to duopoly or cross-ownership rules if the population of the city in question is greater than 400,000.

Concerning allocation and assignment of new public safety services, the Act directed the FCC to reallocate 24 MHZ between TV channels 60-69 for public safety services and to auction the other 36 MHZ in that band for commercial use. The public safety licenses must be assigned in 1998 and the auction must start by January 1, 2001. During the transition to DTV, the FCC must ensure that new spectrum users and existing television licensees could operate without interfering with each other. The bill directs the FCC to seek to assure that qualifying low power TV stations are reassigned other spectrum where possible.

Furthermore, the Act directed the FCC to allocate spectrum for "flexible use," which means defining new services broadly so that services can change as the telecommunications technology evolves. The FCC was already making such allocations, such as allowing specialized mobile radio services to compete with cellular telephone services, or allowing LMDS to provide two-way communications as well as broadcast services. These allocations must be consistent with international agreements, must be required by public safety allocations and in the public interest, and must not result in harmful interference among users.

Shortly after enactment, the FCC made plans for the required auctions, and later conducted a proceeding to determine which wireless services should be exempted from auctions, to determine the appropriate licensing scheme for new and existing services, and to determine how to implement auctions for services that are auctionable as a result of its revised authority.[19]

Status of Spectrum Auctions

To date, the FCC has garnered over $40 billion in total bids from auctioning over 14,300 licenses. According to the FY2002 Bush Administration budget proposal, actual and expected cash receipts total over $31 billion, and spectrum auctions are expected to generate more than $25 billion over the next five years.[20] **Appendix 1** lists types of licenses auctioned to date and an estimate of the expected revenue. As shown, the FCC has conducted auctions for a wide variety of licenses, located at different parts of the radio spectrum, having differing amounts of spectrum, and covering differing geographic ranges. The amounts paid for the licenses depends on these factors as well as many others.

---

[19] FCC 99-52, TW Docket No. 99-87, RM-9332, RM-9405, Notice of Proposed Rule Making in the Matter of Implementation of Sections 309(j) and 337 of the Communications Act of 1934 as Amended, released March 25, 1999.

[20] FCC rules originally allowed winning bidders to make payments in installments according to license terms. Some auction winners, however, defaulted on their payments, causing a decrease in collections. Installment payments are no longer allowed. See: Congressional Budget Office, The Budget and Economic Outlook: Fiscal Years 2001-2010, Appendix B, CBO Baseline for Spectrum Auction Receipts. January 2000. Also, see: CBO, The Budget and Economic Outlook: Fiscal Years 2002-2011, Box 4-1, January 2001.

One important measure of the effectiveness of a licensing scheme is the speed with which licenses are granted. Auctions have proven to be far speedier than either comparative hearings or lotteries, cutting the time required to obtain a license from up to four years to under six months. Although auctions have been fraught with a number of problems, few, if anyone, in government, industry, or academia has advocated returning to the use of comparative hearings or lotteries to assign spectrum licenses.

Most observers consider the auctions to be a success, for the federal revenue generated, as well as for the speed with which licenses auctioned have gone to the companies that value them the most and are most likely to put them to use. Moreover, many prefer letting businesses determine whether to invest in a new service rather than relying on the government to decide who receives a spectrum license. The FCC has concluded that auctioning of spectrum licenses has contributed to the rapid deployment of new wireless technologies, increased competition in the marketplace, and encouraged participation of small businesses.[21] Many other countries have adopted the use of auctions to assign commercial licenses to use spectrum bands.

Recently Completed and Future Scheduled Auctions[22]

The FCC plans to use auctions to assign licenses for the following other wireless services (summary provided in **Appendix 2**):

PCS C & F Block Re-auction
These are the reclaimed licenses from the defaulted licensees in 1996 C- and F-Block auctions. The FCC made several revisions to the service and auction rules, including reconfiguring the size of the C-Block licenses, modifying auction eligibility restrictions, and retaining the spectrum cap on current spectrum license holders. The licenses are in the frequency range from 1890-1975 MHZ. Two 15 MHZ licenses (paired) and four 10 MHZ licenses (paired) are being auctioned in 196 basic trading area (BTA) markets where licenses were reclaimed, for a total of 422 licenses. Some licenses were open to all bidders, while other licenses were available only to small businesses known as entrepreneurs in "closed" bidding. Some, however, have questioned the FCC's rules that allow small company bidders in the current C-Block auction to obtain funding from large companies, arguing that it fails to help small companies enter the wireless industry.[23] The licenses will be used for broadband PCS, which includes a variety of mobile telecommunications services. The auction began on December 12 and ended on January 26, 2001, raising $16.86 billion in revenue, which is more than any single previous FCC spectrum auction. However, the re-auction was thrown into doubt by the June 22, 2001 decision by the U.S. Court of Appeals of the District of Columbia that the FCC did not have the right to reclaim licenses from firms in bankruptcy (i.e., NextWave), and that those licenses still belong to those licensees despite defaulting on their payments.

---

[21] FCC 97-353, FCC Report to Congress on Spectrum Auctions, WT Docket No. 97-150, released October 9, 1997.
[22] See [http://www.fcc.gov/wtb/auctions] for additional details on upcoming auctions.

700 MHZ Guard Band

This spectrum, located in the 746-764 MHZ and 776-794 MHZ bands, was part of the reallocation from television channels 60-69 in connection with the transition to digital television (DTV).[24] To protect public safety users in adjacent spectrum from interference by the new service the FCC established two 6 MHZ "guard bands" at the upper and lower ends of the spectrum allocated for auction. Technical and operational restraints on the use of the guard band spectrum are more stringent than restraints on the other 30 MHZ to be auctioned (discussed below). Auctions for licenses in the guard bands are being conducted in two parts. The first part was conducted in September 2000, raising $520 million. The second part (auctioning eight licenses that were not sold in the first guard band auction) was concluded on February 21, 2001 and raised $20.9 million.

700 MHZ Band

The FY2000 Defense Appropriations Act (P.L. 106-79, Title VIII, Sec. 8124) directed the FCC to conduct auctions for licenses in the upper 700 MHZ band (spectrum reallocated from channel 60-69 television services) so that proceeds are deposited in the U.S. Treasury by September 30, 2000. The FCC originally scheduled auctions for this spectrum (including the guard bands) to meet that deadline. Later, however, the FCC requested permission from House and Senate Appropriations Committee Chairmen to delay the auctions to allow bidders to develop "better business plans and bidding strategies and to form strategic alliances." Although no legislation was introduced, support was informally expressed by the Committees of jurisdiction, and the FCC has most recently postponed most of the upper 700 MHZ band auctions (all but the guard bands) until September 2001.[25]

Spectrum auctioned will be 746-764 MHZ and 776-794 MHZ bands. A total of 12 licenses will be auctioned – one 20 MHZ license (consisting of paired 10 MHZ blocks) and one 10 MHZ license (consisting of paired 5 MHZ blocks) – in six regions known as economic area groupings. These licenses would be considered highly desirable licenses because of their VHF frequency range, except that incumbent television broadcasters are currently using the spectrum, and will continue using it until at least December 31, 2006 (most likely longer). License winners may not cause interference to incumbent broadcasters, making it very difficult to use the spectrum for some time in the more populated parts of the country. The licenses are intended to be used for high-speed Internet access, new fixed wireless operations in under-served areas, and next-generation high-speed mobile services. Accordingly, incumbent channel 60-69 broadcasters have been negotiating a "buyout" with the wireless industry, whereby broadcasters will agree to vacate the spectrum early in exchange for a percentage of the amount wireless companies are expected to pay in auction.[26]

---

[23] Big Carriers Dominate Cellular Auction, Despite Plan to Help Start-Ups, Wall Street Journal, January 5, 2001, p. B1.

[24] Part of this reallocation (the 6 MHz guard bands) was already auctioned in September 2000; the other 24 MHz in those band was reallocated for public safety services.

[25] FCC Memorandum Opinion In the Matter of Cellular Telecommunications Industry Association et al.'s Request for Delay of the Auction of Licenses..., released September 12, 2000.

[26] As a condition for the buyout, some broadcasters are also asking the FCC to require cable companies to carry all of a broadcaster's DTV signals ("multiple carriage"). See: McConnell, Bill, "Paxson Circles the U's," Broadcasting & Cable, February 26, 2001, p. 13.

The FCC has scheduled the upper 700 MHZ band auction for September 12, 2001. The Bush Administration's FY2002 budget proposes legislation that would promote clearing channels 60-69 spectrum for new wireless services in a manner that ensure incumbent broadcasters are fairly compensated. The FY2002 budget proposal would delay the auction until September 2004 in order to increase the spectrum's value and garner more revenue for the Treasury. To date, the FCC has not announced any intention to further delay the auction.

Meanwhile, in a separate proceeding, on March 28, 2001, the FCC issued a notice proposing the auction of the lower 700 MHZ band (698-746 MHZ, television channels 52-59) for commercial wireless services. Public comments on this proposal are due to the FCC by May 14, 2001. The Bush Administration has proposed delaying this auction from 2002 to 2006.

FM Broadcast

These auctions will be for construction permits for FM broadcast stations at 352 locations across the country. Licenses are for the use of spectrum in the FM band (88-108 MHZ) and entail the normal interference protection constraints of FM licenses. The auction is scheduled to begin on December 5, 2001.

Paging Services

The FCC has announced plans to auction 14,000 licenses in the lower paging bands (35-36 MHZ, 43-44 HMZ, 152-159 MHZ, and 454-460 MHZ), and 1,514 licenses in the upper paging bands (929-931 MHZ) that remained unsold in the first paging auction of March 2000. Auction of the lower and upper paging bands is scheduled to commence on June 26, 2001.

Auctions Not Yet Scheduled

The FCC has begun proceedings to plan for auctions for the following other spectrum licenses: AM broadcasting licenses for applicants that filed within designated time frames, licenses for fixed wireless services in the 24 GHz band (called digital electronic message service, or DEMS),[27] licenses around 4.9 GHz transferred from federal government to private sector use, additional licenses for narrowband PCS, low power television (LPTV) and translator stations, and the two additional services described below.

218-219 MHZ (formerly IVDS phase 2)

The FCC had planned to offer two IVDS licenses in each of the 428 rural service areas (RSAs), plus 127 MSA licenses on which the previous winning IVDS bidders defaulted. Just prior to the scheduled start date (February 18, 1997), the auction was postponed indefinitely as a result of numerous petitions from industry and Congress to revise the service rules to make these licenses more attractive to bidders. In 1999, the FCC offered financial restructuring to current IVDS licensees, and debt forgiveness to previous licensees who made their first two down payments but did not make their March 16, 1998, payment. The FCC also

---

[27] DEMS licenses will each be 40 MHz (paired).

changed the IVDS service rules to increase the flexibility of licenses to allow the provi9son of Internet services.

General Wireless Communications Services (GWCS)

The FCC has allocated five licenses, 5 MHZ each, between 4660-4685 MHZ, to be auctioned in each of 175 Economic Areas, for a total of 875 licenses. There are, however, incumbent licenses in that spectrum in many parts of the country. GWCS licenses, conceived to be similar to WCS, may be used to provide any fixed or mobile communications services except broadcast, radiolocation, and satellite services. These may include voice, video, and data services, private microwave, broadcast auxiliary, and ground-to-air signals. The question of what to do with the incumbents, however, is causing difficulty in designing the auction.

Spectrum Value

Spectrum value depends on many factors, such as the amount of spectrum, its frequencies (since signal transmission characteristics vary along different parts of the spectrum), the geographic area covered, the services permitted by FCC rules, the availability of equipment that can operate at those frequencies, the demand for services that do not interfere with other bands, the amount of alternative spectrum already available for similar services, the number of incumbents presently occupying the spectrum, and whether incumbents will remain in that spectrum or be relocated to other spectrum. Spectrum value may be greater if adjacent bands can be aggregated to form larger blocks and if the given spectrum is not encumbered by other licensees using the same frequencies. Giving bidders enough time to review the auction rules and services rules, examine technical opportunities and constraints, prepare marketing plans, and arrange financing is also critical to obtaining full value of the auction. It is impossible to determine in advance precisely the revenue that can be obtained from a given spectrum auction.

After an auction closes, spectrum value is often measured by the total dollars raised per "MHZ-pop" (the number of MHZ provided in a license multiplied by the total population covered by the license, similar to a unit price). However, the MHZ-pop value of a given license can vary significantly from one auction to another. In the PCS auctions, for example, the narrowband PCL licenses drew over six times more revenue per MHZ-pop than the broadband PCS licenses, but drew much less total revenue because of the smaller amount of spectrum auctioned.

The Congressional Budget Office (CBO) annually scores the anticipated receipts from planned spectrum auctions, and includes the revenue estimate in its annual report, *The Budget and Economic Outlook*.[28] The January 2001 report estimates receipts from the spectrum at nearly $28 billion over the 2001 through 2011 period. The revenue expected from the auctions is used as offsetting receipts to other federal expenditures. In accordance with the Budget Enforcement Act of 1990,[29] the auction proceeds, as assessed by CBO, cannot be used for funding other programs.

---

[28] CBO does not break down the amounts in terms of individual auctions, but only provides the aggregate estimated revenue.

[29] For a discussion of the Budget Enforcement Act of 1990 and other budgetary requirements, see CRS Report RL30363, The Sequestration Process and Across the Board Spending Cuts for FY2001.

Technology Innovations

Several technological advances could affect the outcome and prospects for spectrum auctions and how the spectrum is managed. The usable spectrum for communications purposes is currently considered to be below 300 GHz. Higher frequencies present limitations such as a greater absorption of signals by the atmosphere, and difficulties in high frequency reception. As the technology for radio transmission and reception improves, higher frequencies will likely become available for use. Technology improvements may, in turn, spur increased consumer demand for spectrum.

Some of the problems with high-frequency signal transmission and signal interference at all frequencies are being solved by engineering techniques which could make better use of the spectrum, thus reducing some of the spectrum demand. These include methods of digital signal compression, which increases the carrying capacity of currently sued bands, error detection and correction which maintain the signal integrity even in high levels of noise, and other digital techniques such as frequency hopping, in which the transmitted signal avoids frequencies that are already being used. The use of fiber optic cables (which carry signals over wires rather than propagating through the air, and therefore do not require frequency allocations) can provide enormous capacity and alleviate some of the demand for spectrum. Cables, of course, cannot be used for services that require wireless transmission. Two other advanced technologies called ultra-wide band technology and software-defined radio, are discussed under **Recent Developments in Spectrum Management**.

In developing service and auction rules for spectrum license auctions, the FCC tries to maximize spectrum flexibility by allowing licensees to offer as many services as possible without interfering with existing spectrum use. The FCC also usually considers competitive market forces in allocating and licensing spectrum. This entails using auctions for many new terrestrial spectrum licenses, and defining new services broadly enough to allow services to change as the technology evolves.

In some auctions, spectrum is sold for the same spectrum bands in the same geographic areas as incumbent licensees are located. The new licenses are called overlay licenses because they use frequencies that surround the frequency of an existing license. The auction winner must prevent the operations of its overlay license from interfering with those of an incumbent licensee. The new licensee could either "work around" the spectrum of the incumbent license (by using frequency hopping) or pay for the relocation of the incumbent to some other frequency. Overlay licensees were implemented in the PCS auctions since there were already incumbent licensees (called microwave licensees) using that spectrum. To help clear the PCS spectrum of microwave incumbents, the FCC provided higher frequency spectrum for the incumbents and required PCS licensees to pay for the costs of relocating incumbents to high frequencies. However, placing that requirement on the new licensee typically lowers the value of the license. Most licenses currently auctioned by the FCC are encumbered with existing licensees to an even greater extent that the PCS spectrum. If the cost of relocating the incumbent exceeds the value of the license, it can be difficult to attract bidders in an auction. A further difficulty is that in some auctions, the FCC has not provided new spectrum to relocate incumbents, leaving those negotiations for after the auction.

To help smaller businesses participate in an auction, the FCC sometimes allows license winners to partition licenses into smaller geographic areas than were originally defined by the FCC. This allows a wireless service provider to set up a business in a smaller community without having to serve an entire region. The FCC also allows some licensees to "disaggregate" a portion of their spectrum, i.e., to divide the spectrum into several smaller bands. This enables smaller companies to use a portion of the spectrum for some specialized service.

Recent Developments in Spectrum Management

Intelligent Transportation Systems

The automotive industry and the federal government have been working together for many years developing the electronics, communication systems, and information processing capability to improve the efficiency and safety of surface transportation systems. These planned systems are collectively referred to an intelligent transportation systems (ITS).[30] In October 1999, in accordance with provisions in the Transportation Equity Act for the 21st Century (23 U.S.C. 502),[31] the FCC allocated 75 MHZ in the 5.850-5.925 GHz band for ITS on a co-primary basis with current users of that spectrum. ITS users were given the same level of protection from interference as other primary users (which include federal (primary defense) operations, and commercial fixed satellite services), and a higher level of protection than secondary users (amateur radio). ITS will use this spectrum for dedicated short range communications such as traffic light control, traffic monitoring, travelers' alerts, automatic toll collection, and other purposes. The FCC deferred consideration of licensing and service rules to a later proceeding, in anticipation of further details on ITS requirements from the Department of Transportation.

Technology Advisory Council

In early 1999, with the increasing number of conflicts associated with spectrum management, and the burgeoning wireless communications industry, the FCC established the Technology Advisory Council. The Council, comprised of industry and academic experts, would help the FCC in planning its strategy for regulating the wireless industry, as well as analyze issues of convergence with the Internet and other technical issues. Since its inception, the Council has studied issues of spectrum management, electromagnetic noise created by interference among commercial and government wireless systems, access to telecommunications by persons with disabilities, network interconnection, and network access.

The Council investigated two technologies in particular, which are thought to have the potential to alleviate some of the demand for spectrum. One, called *ultra wide-band* technology, is an innovation involving the spreading of a radio signal over a wide range of frequencies that are already assigned to other communications services. If the ultra wide-band signal is transmitted at a low enough power level, it would not interfere with existing signals. The other technology, called *software-defined radio*, is the development of a new type of

---

[30] For information on ITS, see CRS Report RL30403, Intelligent Transportation Systems: Overview of the Federally Supported Research and Development Program, January 11, 2000.

[31] The provision directs the Secretary of Transportation, in consultation with the Secretaries of Commerce and Defense and the FCC, to secure the necessary spectrum for the establishment of dedicated short-range vehicle to wayside wireless services.

radio equipment that can be quickly reprogrammed to transmit and receive on any frequency within a wide range using any transmission format. The FCC has proposed allowing the limited use of both of these technologies on an unlicensed basis. However, incumbent users of the spectrum that would be used for these technologies are concerned about the FCC proposal. The incumbents (primarily the cellular and PCs industries, and operators and users of the Global Positioning System satellite navigation system) argue that potential interference may cause failures in their systems, some of which involve public safety.

FCC Policy Statement and Regulatory Framework for Auctions

In November 1999, the FCC released a statement outlining guiding principles for its future activities in spectrum management.[32] The statement was intended to provide a framework for industry and government parties to understand future FCC decisions regarding reallocation of the remaining 200 MHZ of spectrum to be transferred from government to non-government radio services as per statutory requirements. The new strategy contained the following principles:

- allow greater flexibility in allocations, including "harmonization" of FCC service rules to provide regulatory neutrality for similar wireless services;

- promote new spectrum-efficient technologies, such as ultra-wideband and spread spectrum operations;

- ensure that important communications needs, such as public safety, are met;

- improve the efficiency of assigning licenses through streamlining and innovative techniques;

- encourage the development of secondary markets for spectrum (i.e., reselling of licenses to third parties) to ensure full utilization; and

- seek ways to make more spectrum available, through, for example, assigning user fees or by reclaiming existing spectrum.

One innovation that was introduced to improve the efficiency of assigning licenses is the "band manager" concept. Under this approach, licenses for blocks of spectrum would be auctioned to band managers, who would then subdivide and lease portions of their spectrum in response to market demand. Prices charged to users would be set by competition among the band managers for potential spectrum users.

The Policy Statement then inventoried spectrum that was available for allocation and outlined proposals for allocating and assigning that spectrum. At the same time, the FCC created a Spectrum Policy Executive Committee to address policy issues affecting spectrum management, to implement initiatives consistent with the Policy Statement, and to coordinate related actions among the FCC's internal bureaus.

---

[32] FCC 99-354, Policy Statement. In the Matter of Principles for Reallocation of Spectrum to Encourage the Development of Telecommunications Technologies for the New Millennium, released November 22, 1999.

In November 2000, the FCC established a regulatory framework for future auctions, which provided further details on its spectrum management plans, and answered many questions raised in the proceeding it initiated in March 1999.[33] In the framework, the FCC may conduct auctions for licenses for private radio services to resolve mutually exclusive applications if the FCC "determines that it is in the public interest to do so." The FCC will continue to decide on a service-by-service basis the licensing scheme for new services. One option discussed is to use a "band manager" concept (described above) which has been used in the 700 MHZ Guard Band auction in September 2000. In addition, the ruling defines the scope of the statutory exemption from auction for public safety radio services to include not only police, fire, and emergency medical services, but also non-commercial services used by entities such as utilities, railroads, and transit systems. The ruling also addressed a number of pending petitions to amend its licensing and eligibility rules for existing private wireless services (see **Issues for Congress, Private Land Mobile Radio Services**).

Recent Spectrum Allocations

In October 2000, the FCC allocated 50 MHZ (3650-3700 MHZ, which had previously been transferred from government use) for commercial wireless services.[34] The FCC intends to permit and encourage the use of this spectrum for new broadband, high-speed wireless voice and data services, particularly in rural areas. The spectrum will continue to be encumbered by existing licensees in the fixed satellite services, which will share the band with new licensees. The FCC proposed to assign the new licenses through auctions.

In November 2000, the FCC proposed to reallocated 27 MHZ of spectrum, previously transferred from federal government use, to various non-government services.[35] The spectrum, located in a number of separate bands, could be allocated for private land mobile services, satellite feeder links, utility telemetry to support meter reading, and personal location services. Some of the licenses could be assigned to band managers through auctions.

In another November 2000 action, the FCC proposed to adopt a new policy to promote the development of "secondary markets" in radio spectrum.[36] The FCC Policy Statement articulates its goal of promoting a system of secondary markets (i.e., the sale and lease of the right to use spectrum by licensees) to better utilize spectrum that is already licensed. The FCC proposed to allow wireless radio services licensees, with exclusive rights to their assigned spectrum, to lease their spectrum rights to third parties without having to secure prior FCC approval.

In January 20001, the FCC released a proposal to consider possible uses of several frequency bands below 3 GHz for new advanced wireless systems, including third generation

[33] FCC 00-403, WT Docket No. 99-87, Report and Order and Further Notice of Proposed Rule Making (NPRM), released November 20, 2000.

[34] FCC 00-363, Docket No. 98-237, First Report and Order and Second NPRM, announced October 12, 2000.

[35] FCC 00-395, ET Docket no. 00-221, NPRM In the Matter of Reallocation of...Government Transfer Bands, released November 20, 2000.

[36] FCC 00-041, Policy Statement, released December 1, 2000, and FCC 00-402, WT Docket 00-230, NPRM Promoting Efficient Use of Spectrum Through Elimination of Barriers to the Development of Secondary Markets, released November 27, 2000.

(3G) systems (see discussion of 3G spectrum).[37] Portions of the 1710-1850 MHZ and 2110-2165 MHZ bands (previously transferred from federal government to non-government use) could be allocated for 3G services, and various approaches for using the 2500-2690 MHZ band will be considered. In the same action, the FCC adopted an Order denying a petition by the Satellite Industry Association for parts of the 2500-2690 band to be reallocated to mobile-satellite services.

ISSUES FOR CONGRESSIONAL CONSIDERATION

Allocation of Spectrum for Federal vs. Commercial Use

Tension, which has always existed between federal agencies and the private sector over spectrum allocations, has been increasing in recent years. The Omnibus Budget Reconciliation Act of 1993 (47 U.S.C. 927) directed federal agencies to vacate 200 MHZ of spectrum below 3 GHz for reassignment to commercial uses. NITA, manager of all spectrum used by the federal government, identified 235 MHZ to transfer to the FCC,[38] but claimed that releasing any additional spectrum could result in costs greater than the potential revenue from an auction, and could compromise national security, public safety, law enforcement, and air traffic control operations.[39] Nevertheless, a provision in the FY1997 Omnibus Appropriations Act (P.L. 104-208) directed the FCC to auction 30 MHZ of spectrum previously allocated for shared commercial and government radio services (included as part of the 235 MHZ identified by NTIA). The Department of Defense (DOD), the largest federal user of spectrum, was the most vocal of the federal agencies protesting the transfer of spectrum away from federal use. DOD argued that it needs all of the spectrum it is currently assigned to maintain high quality communications to support national security.

Amid the protests by NTIA, DOD, and other federal agencies, the commercial sector increased its pressure on Congress to release additional federal spectrum for commercial use. The Balanced Budget Act of 1997 (P.L. 105-33) directed NTIA to reassign an additional 20 MHZ below 3 GHz to the FCC for auction. In 1998, NTIA identified 20 MHZ (located in five spectrum bands) and a schedule for reallocation from federal agency use within ten years. The report concluded, however, that such a spectrum release could adversely affect critical agency missions and the ability to provide services to the public.[40] The report estimated the cost to federal agencies to be over $1 billion to modify existing equipment and facilities to use alternative frequencies, assuming that suitable spectrum will be available, based on the assumption that extensive system modifications would not be required to avoid interfering with new commercial users. The report stated that the loss of the identified spectrum could

---

[37] FCC 00-455, ET Docket No. 00-258, NPRM and Order In the Matter of Amendment of Part 2 of the Commission's Rules to Allocate Spectrum Below 3 GHz for Mobile and Fixed Services to Support..., released January 5, 2001.

[38] The amount of spectrum identified by NITA was 35 MHz greater than required. This was possibly in anticipation of future demand for commercial spectrum.

[39] Testimony of Hon. Larry Irving to House Commerce Committee, Subcommittee on Telecommunications, Finance, March 21, 1996, and to Senate Commerce Committee, June 20, 1996.

[40] NTIA Publication 98-36, Spectrum Reallocation Report: Response to Title III of the Balanced Budget Act of 1997, released February 1998.

restrict spectrum use during defense training exercises, ultimately affecting operational readiness and national security.

In two follow-on reports, NTIA warned that reallocating this spectrum would have a "profoundly negative impact on the planned U.S. space program" and other federal systems, and identified alternative bands to be considered for auction and use for new wireless services.[41] The alternate spectrum included non-federal bands, bands that NTIA had previously released to the FCC, and a shared band between federal and non-federal uses. The telecommunications industry, criticized the NTIA's selection, claiming that the alternative bands were undesirable for commercial use and were encumbered by federal users. In October 2000, the FCC reallocated the bands recommended by NTIA for commercial services, and proposed to assign the licenses through the use of auctions.[42]

While the FCC has reallocated many of the bands identified by NTIA for transfer to non-federal uses, several bands must still be reallocated and licensed. **Table 1** provides a list of the spectrum bands identified by NTIA for transfer to commercial uses, but which either have not yet been reallocated or have not yet been licensed to commercial services. The table indicates the current allocations for each of the bands, the status of the transfer of the bands to non-federal uses, and statutory auction deadlines, where applicable. Congress will likely continue to be pressured by the wireless industry to expedite the FCC's reallocation and licensing of these bands for commercial use. Another issue in which Congress could become involved is how to compensate federal agencies for relocating their wireless services to new frequencies. NTIA is planning to begin a proceeding on that issue.

In addition, two legislative measures enacted in recent years will affect the transfer of spectrum from federal to non-federal uses. The FY1999 Defense Authorization Act (47 U.S.C. 923, Title X, Sec. 1064) requires any entity that purchases a license for spectrum previously reserved for used by a federal agency to reimburse that agency for the costs incurred by the agency in relocating its communications to other frequencies. Previously reallocated spectrum and reallocations n the 1710-1755 MHZ band are exempt from this reimbursement. This provision will have the effect of lowering the value of the spectrum at auction, and could cause delays in the licensing process if there are disputes between federal users and license winners over the costs of relocation.

---

[41] NTIA Publication 98-37, Reallocation Impact Study of the 1990-2110 MHz Band, and NTIA Publication 98-39, Identification of Alternate Bands: Response to Title III of the Balanced Budget Act of 1997, both released November 1998.
[42] FCC 00-363, Docket No. 98-237, First Report and Order and Second NPRM, released October 24, 2000.

Table 1. Spectrum to be Reallocated from Federal to Non-Federal Uses

Spectrum Band (MHZ)	Primary Current Allocations	Status of Reallocation	Statutory Auction Deadline
216-220	Maritime mobile, inter-active video and data service (218-219 MHZ)	FCC Rulemaking in progress (FCC 00-395)	9/30/2002 (BBA97)
1390-1395	Fixed, radiolocation	FCC Rulemaking in progress (FCC 00-395)	None
1395-1400	Wireless Medical Telemetry Service	Reallocated by FCC to Wireless Medical Telemetry Service*	None
1427-1429	Space operation, fixed mobile	FCC Rulemaking in progress (FCC 00-395)	None
1429-1432	Wireless Medical Telemetry Service	Reallocated by FCC to Wireless Medical Telemetry Service* FCC Rulemaking in progress (FCC 00-395)	None
1432-1435	Fixed, mobile	FCC Rulemaking in progress (FCC 00-395)	9/30/2002 BBA97)
1670-1675	Meteorological satellites, meteorological aids	FCC Rulemaking in progress (FCC 00-395)	None
1710-1755	Fixed, mobile	FCC Rulemaking in progress (FCC 00-455) Released January 5, 2001	9/30/2002 (BBA97)
2300-2305	None	FCC set aside as reserve spectrum	None
2385-2390	Mobile, radiolocation	FCC Rulemaking in progress (FCC 00-395)	9/30/2002 (BBA97)
2400-2402 2417-2435	None	FCC set aside as reserve spectrum	None
3650-3700	Fixed satellite, radio-location, aeronautical radionavigation	FCC Rulemaking in progress (FCC 00-363)	None
4940-4990	Fixed, mobile	Substituted by the President for the originally identified 4635-4685 MHZ band. FCC Rulemaking in progress (FCC 00-363)	None

Source: CRS, based on FCC and NTIA data

OBRA93=Omnibus Budget Reconciliation Act of 1993, BBA97=Balanced Budget Act of 1997

* These bands will be shared on a co-primary basis with existing government operations (FCC 00-211 June 2000).

The FY2000 Defense Authorization Act (P.L. 106-65, Title X, Sec. 1062, *Assessment of Electromagnetic Spectrum Reallocation*), directed the Department of Commerce (DOC) and the FCC to conduct a review and assessment of national spectrum planning; the reallocation of federal spectrum to non-federal use in accordance with existing statutes; and the implications for such reallocations to the affected federal agencies. Particular attention was to be given to the effect of the reallocations on critical military and intelligence capabilities, civilian space programs, and other federal systems used to protect public safety, as well as future spectrum requirements of federal agencies. In response to this requirement, as part of the review and assessment of spectrum planning, in November 2000 the FCC requested comments from industry on procedures for reimbursement of relocation costs to federal spectrum users.[43] In January 2001, NTIA proposed rules governing reimbursement to federal entities by the private sector related to reallocation of spectrum.[44] Final rulings by both the FCC and NTIA are expected later this year.

The FY2000 Defense Authorization Act further states the DOD is not required to transfer any spectrum bands to the FCC unless NTIA, in consultation with the FCC, makes available an alternative band or bands to DOD. Further, DOD, DOC, and the Joint Chiefs of Staff must certify that the alternative band (or bands) provides comparable technical characteristics to maintain essential military capability. The Act further requires that 8 MHZ (located in the following three bands: 139-140.5 MHZ, 141.5-143 MHZ, and 1385-1390 MHZ) that were identified for auction in the Balanced Budget Act of 1997, be reassigned to the federal government for primary use by DOD. The conference report (H.Rept. 106-301) urges DOD to "share such frequencies with state and local public safety radio services, to the extent that sharing will not result in harmful interference between DOD systems and the public safety systems proposed for operation on those frequencies." Those 8 MHZ were subsequently reclaimed by the President for exclusive federal use.

The provision in the FY2000 Defense Authorization Act reclaiming 8 MHZ of spectrum for DOD use had an impact on two reallocation proceedings. First, the FCC was developing a plan for the use of part of the 8 MHZ (the 138-144 MHZ band) for interoperable communications among federal, state, and local public safety wireless systems. Another band of spectrum will have to be found for that purpose.[45] Second, in February 1998, in fulfillment of the Balanced Budget Act of 1997, NTIA identified another part of the 8 MHZ (139-143 MHZ) for the FCC to reallocate and assign by auctions. Since the FY2000 Defense Authorization Act supercedes previous laws, that spectrum will not be transferred, causing a shortfall in expected revenue from spectrum auctions in FY2002. The Congressional Budget Office (CBO), however, estimated the budget impact of foregone spectrum receipts due to this provision to be $500 million or less. This low estimate was due, in part, to the requirement that spectrum auction winners reimburse the federal agencies for the costs associated with relocating to new frequencies. Congress might decide to monitor the

---

[43] FCC 00-395, ET Docket No. 00-221, NPRM In the Matter of Reallocation of Government Transfer Bands, released November 20, 2000.

[44] NTIA Docket No. 001206341-01, RIN 0660-AA14, Notice of Proposed Rule Making posted in Federal Register January 18, 2001.

[45] FCC WT Docket 98-86, First Report and Order and Third NPRM, Development of Operational, Technical, and Spectrum Requirements for Meeting Federal, State, and Local Public Safety Agency Communication Requirements Through the Year 2010, released September 29, 1998.

implications of these laws and related actions on future reallocations of spectrum for federal or commercial uses. The General Accounting Office is currently investigating this issue.

Spectrum Auction and License Payment Procedures

As a result of a number of problems associated with the spectrum auctions (in particular the C-Block auction), in October 1997 the FCC recommended to Congress the following legislative actions:[46]

- to clarify that FCC licensees who default on their installment payments may not use bankruptcy litigation to refuse to relinquish their spectrum licenses for re-auction;

- to grant the FCC explicit statutory authority to manage its installment payment portfolio flexibly;

- to exempt auction contracts from certain provisions of the Federal Acquisitions Regulations (FAR); and

- to modify the statue of limitations for forfeiture proceedings against non-broadcast licensees from one to three years.

In addition, the FCC streamlined its rules to simplify the auction process (e.g., the applications and payment procedures for bidders), and adopted uniform affiliation rules and ownership disclosure rules to avoid legal problems associated with a 1995 Supreme Court decision limiting special treatment for women-owned, and minority-owned companies.[47] The revised rules also provide for higher bidding credits for small businesses (15, 25, and 35 percent, based on the size of the business).

Legislation was introduced in 1998, and again in 1999, addressing these concerns. The 1999 provision was in a section of the Senate version of the FY2000 Appropriations bill for the Departments of Commerce, Justice, State, and Related Agencies (S. 1217, Sec. 618, introduced June 14, 1999) that would have met some of the FCC's requests. The provision authorized the FCC to recover and re-auction licenses if a license fails to meet its installment payment obligations; it allowed the FCC to avoid the jurisdictional and administrative burden associated with reclaiming a license under state laws; and the provision was retroactive to include pending cases. Several wireless service providers opposed the provision, and it was removed in conference.

The provision on auction procedures was not included in FY2001 Appropriations or any other legislation introduced in 2000. Some of the motivation for the legislation has subsided since the FCC now requires payment in full by license winners. Moreover, a series of court decisions over the past several years involving the defaulted C-block licensees has enabled the FCC to proceed with the re-auction of these licenses. However, in the unlikely event that a

---

[46] FCC 97-353, FCC Report to Congress on Spectrum Auctions. WT Docket No. 97-150, released October 9, 1997.

[47] See Footnote 13.

higher court overturns the latest ruling, winners of some of these licenses will have to give them back.[48]

The FCC continues to express a need for a provision to establish its regulatory authority over spectrum licenses in all states and jurisdictions in the country.[49] Some companies in the wireless industry continue to oppose such a provision, while other companies advocate it. An issue before the 107th Congress is whether to review the FCC's recommendations in light of all of the recent changes made in spectrum management policy to determine whether a legislative remedy is warranted.

Digital Television Spectrum

Digital television (DTV) is a new television broadcasting service developed by the television and computer industries. Without the authority to conduct auctions for DTV licenses, the FCC granted, free of charge, licenses to broadcasters for DTV transmissions, as directed by the Telecommunications Act of 1996 (47 U.S.C. 336). In April 1997 the FCC granted all full power television broadcasters (over 1600 stations), as part of their licenses, 6 MHZ to transmit DTV programming over the air in addition to their 6 MHZ of analog television spectrum.[50] The FCC and the television industry are planning a transition from the current analog television broadcasting system to DTV over the next several years. The FCC's plan is that at the end of the transition to DTV, the broadcasters will return the 6 MHZ currently used for analog television broadcasting. Some in industry, the FCC, and Congress are concerned that the transition is not proceeding rapidly enough, and that DTV might fail to become competitive with other multi-channel video services.

During the transition to digital television, broadcasters are transmitting both the existing analog television and DTV signals, so that consumers can continue to receive television programming on their existing receivers until they purchase new digital television sets or converters. Television stations in the nation's top ten markets started to provide a digital signal on November 1, 1998. All remaining commercial DTV stations are to be constructed by May 1, 2002, and non-commercial DTV stations by May 1, 2003. The FCC had originally set 2006 as the target date for broadcasters to cease transmitting the analog signal and return the 6 MHZ of analog television spectrum to the FCC to use for other purposes. The Balanced Budget Act of 1997 (P.L. 105-33), however, prohibited the FCC from reclaiming the analog spectrum from a broadcaster if at least 15 percent of the television households in that broadcaster's market have not purchased DTV receivers or converters. Most analysts believe it is unlikely that most markets will have met that criterion by the 2006 target date. Broadcasters in some markets may not have to surrender their spectrum for many more years. Some question whether the FCC will ever be able to reclaim the spectrum. In aggregate, the auctionable analog TV spectrum represents a large amount (about 84 MHZ) of highly

---

[48] E-Business Auction: Airwaves Auction Pulls in 16.86 Billion, Wall Street Journal, January 29, 2001, p. B4.

[49] The court decision giving the FCC the authority to re-auction the licenses to defaulted C-Block licensees only applied to the legal jurisdiction that includes New York, where the defaulted licensees (most notably NextWave Inc.) were headquartered.

[50] FCC Fifth Report and Order on Advanced Television Systems and Their Impact on Existing Television Service, released April 21, 1997.

desirable spectrum.[51] However, the uncertainty over the availability of this spectrum will present difficulties for the FCC in attempting to attract bidders to an auction for the spectrum.

DTV services currently consist only of high definition television (HDTV) during prime-time programming hours, but broadcasters are also planning to use their DTV spectrum during other parts of the day to simultaneously transmit multiple television programs in digital format instead of a single HDTV broadcast. The DTV spectrum can also be used to provide interactive television, Internet access, subscriptions to multimedia news services, and other information services. Each new service provides new options for consumers and new revenue to broadcasters. Currently, over 500 DTV stations have been granted construction permits, and over 180 are on the air. Despite criticism from observers over the slowness of the transition to DTV, the FCC has stated that most of the delays in construction are a result of matters beyond the broadcasters' control.[52]

Several issues could cause delays in the transition to DTV and in the return of the analog television spectrum. One issue is whether Congress should amend the Balanced Budget Act of 1995 to require broadcasters to return the analog spectrum under stricter deadlines. Another concern is over the adequacy of the FCC's rules requiring fees from broadcasters who use DTV licenses for subscription services.[53] Another involves what public interest requirements should be placed on DTV licensees. Controversy continues between state and local authorities and broadcasters (and other wireless service providers) over the placement and construction of DTV transmission towers. The FCC has not yet determined all of the conditions under which cable TV operators should be required to provide the DTV programming of broadcast TV stations (i.e., the "must carry" debate). Disagreements also continue among cable service providers, television manufacturers, and broadcasters over a standard for connecting DTV sets with digital cable systems. Finally, the transition to DTV depends on the willingness of consumers to purchase DTV receivers or converters. For each of these issues, Congress will likely have to decide to what extent, if any, it should intervene in the development of markets and regulations.[54]

Meanwhile, the Bush Administration's FY2002 budget has proposed "squatting fees" as an incentive for broadcasters to surrender their analog spectrum. Under this proposal, the broadcasters would pay the U.S. Treasury $200 million per year in analog spectrum fees from 2002 till 2006. Between 2006 and 2010, fees would be reduced until, it is assumed, all the analog spectrum will have been reclaimed. While similar "squatting fees" were proposed by the previous two Administrations, Congress has not implemented nor endorsed this approach.

Low Power FM Radio Service

In response to numerous inquiries from religious and other local groups, in January 2000, the FCC adopted rules for a low power FM radio (LPFM) broadcasting service to be licensed

---

[51] See Completing the Transition to Digital Television, Congressional Budget Office, September 1999. page 14.

[52] Statement of Dale Hatfield, Chief, FCC Office of Engineering and Technology, before the House Commerce Committee, Subcommittee on Telecommunications, Trade and Consumer Protection, July 25, 2000.

[53] Consumer groups would like the fees to be higher for commercial broadcasters, while the broadcasters think the fees should be lower.

[54] For further analysis of these issues, see CRS Report 97-925, Digital Television: Recent Developments and Congressional Issues, updated February 2, 1999.

to local communities. The service consists of two classes of LPFM radio stations with maximum power levels of 10 watts and 100 watts. The rules contain interference protection criteria to help ensure that the LPFM service protects and preserves the technical integrity of existing radio service. Since the inception of LPFM, however, incumbent full-power FM broadcasters and radio manufacturers have protested the FCC ruling. The main argument against LPFM is based on concerns over interference with existing FM radio broadcasts, and the potential that LPFM might impede the future transition to digital audio broadcasting.

Many Members sought to severely scale back or nullify the FCC's decision to issue LPFM licenses, while other Members supported the FCC ruling. On April 14, 2000, the House passed the Radio Broadcasting Preservation Act of 2000 (H.R. 3439, as amended), to prevent LPFM licenses from being located within three FM channels away from incumbent full-power broadcasters, while directing an independent study of whether LPFM is causing harmful interference to full-power broadcasters. Three bills were introduced in the Senate: S. 2068, to prohibit the FCC from authorizing LPFM licenses; S. 2518 (later introduced in modified form as S.2989), to permit a limited number of LPFM licenses; and S. 3020, similar to the House-passed bill. Language similar to the House-passed bill was inserted into the FY2001 Appropriations Bill for the Departments of Commerce, Justice, and State, and related agencies (H.R. 4690), which was not signed by President Clinton, citing the LPFM provision as one of the issues that must be resolved.[55] The provision was again included in the District of Columbia Appropriations bill (H.R. 4942), (Conf. Rept. 106-1005 contained the Commerce, Justice, and State Appropriations bill, H.R. 5548) that passed Congress on December 15, 2000. It is estimated that the legislation will have the effect of eliminating 75 percent of the potential LPFM licenses that would otherwise be granted.

The 107th Congress might decide to monitor the results of the required study. If it is found that LPFM does not cause harmful interference to full power broadcasters, Congress could reinstate the FCC's original LPFM program.

Third Generation (3G) Mobile Wireless Services

Rapid growth in the number of subscribers of mobile wireless telecommunications services in the United States and abroad is fueling interest and developments in the next generation of wireless technology services, known as 3G. In addition to the existing wireless communications capabilities, 3G services might include high-speed mobile Internet access and the ability to use the same handset anywhere in the world. Issues related to the implementation of 3G centers mainly on the allocation of spectrum and adoption of technology standards by each of the countries developing this new service. While some steps have been taken to coordinate these activities, much work remains before 3G services will be available to the American public.

The International Telecommunication Union (ITU), a United Nations (UN) agency, is sponsoring the adoption of 3G standards and the allocation of spectrum to integrate various

---

[55] For further discussion and analysis of the LPFM issues, see CRS Report RL30462, Low Power FM Radio Service: Regulatory and Congressional Issues, updated regularly.

satellite and terrestrial mobile systems into a globally interoperable service.[56] The ITU conducts World Radiocommunication Conferences (WRCs) every two to three years to reach consensus among member states on spectrum allocations. One of the key issues at WRC-2000 was the identification of global spectrum bands that could meet the additional spectrum requirement for 3G services. Once the spectrum bands for 3G were identified internationally, each country had to decide what frequencies within those bands to use for the initial implementation of 3G services, as well as the long-term expansion of those services.

One of the identified bands, 1755-1850 MHZ, is currently allocated in the United States for exclusive government use. While the EU would like that spectrum to be allocated for 3G services in the United States, some federal agencies, particularly the Department of Defense (DOD), are concerned that any 3G services that are licensed in that band could interfere with existing communications. Similarly, another band identified for 3G at WRC-2000, 2500-2690 MHZ, also has incumbent licenses in the United States. Incumbents include multi-channel multipoint distribution systems (MMDS, a commercial "fixed wireless" service originally used for television broadcasts, and now being developed for mobile wireless broadband applications), and Instructional Television Fixed Services (ITFS, similar to MMDS but licensed for educational programming).

The need to expedite spectrum 3G allocations was highlighted on October 13, 2000, when a Presidential Memorandum issued by President Clinton directed all federal agencies to work with the FCC and the private sector to identify the spectrum needed for 3G services. The FCC, in conjunction with NTIA, was directed to identify suitable 3G spectrum by July 2001, and to auction licenses to competing applicants by September 30, 2002. Both the FCC and NTIA were tasked with conducting studies into the potential for allocating 3G spectrum. NTIA's final report (*The Potential for Accommodating Third Generation Mobile Systems in the 1710-1850 Band*, March 2001) found that full-band sharing is not feasible because of signal interference with DOD systems, and that relocating DOD spectrum to another band would not be possible, if at all, until beyond 2010.[57] The NTIA has suggested that some limited band sharing options might be feasible if 3G operations can be restricted in space or time, and if 3G operators reimburse certain federal operators to relocated to new frequencies prior to commencing operations near those federal operations. On January 18, 2001, NTIA released a Notice of Proposed Rule Making (NPRM) on the reimbursement procedures associated with the use of that band, and will issue a final rule later this year (for further details see NTIA's 3G website [http://www.ntia.doc.gov/ntiahome/threeg/index.html].

The FCC's final report (Spectrum Study of the 2500-2690 MHZ Band, March 30, 2001) found that band sharing between 3G systems and incumbent licensees (primarily schools with instructional television licenses (ITFS) and fixed wireless broadband providers) would be problematic, and that there is no readily identifiable alternate frequency and that could accommodate a substantial relocation of the incumbent operations in the 2500-2690 MHZ band. In December 2000, the FCC released an NPRM proposing to allocate portions of the

---

[56] In mobile terrestrial systems, the individual handsets send and receive the telecommunications signals to and from nearby ground-based stations that connect to the public switched telephone network. In mobile satellite systems, the handsets send and receive signals to and from orbiting satellites that relay the signals to a single ground station that serves a large geographic area, and then connects to the public switched telephone network.

[57] On October 30, 2000, DOD released a report on the electromagnetic compatibility interactions between major DOD radiocommunications systems operating in the 1755-1850 MHz band and potential 3G systems. The report stated that the band is heavily used by government users.

1710-1850 MHZ and 2110-2165 MHZ bands (previously transferred from federal government to non-government use) for 3G services, and seeking comment on various approaches for using the 2500-2690 MHZ band. The FCC also adopted an Order denying a petition by the Satellite Industry Association for parts of the 2500-2690 band to be reallocated to mobile-satellite services (for further details see the FCC's 3G website [http://www.fcc.gov/3G/]).

With the July 2001 deadline approaching for 3G spectrum allocation decisions, on June 26, 2001, FCC Chairman Michael Powell sent a letter to Secretary of Commerce Donald Evans recommending that the deadline be extended "to allow the Commission and Executive Branch to complete a careful and complete evaluation of the various possible options for making additional spectrum available for advanced wireless services." Chairman Powell recommended extension of statutory deadlines as well, stating that "[t]ogether with the Executive Branch and our congressional authorizing and appropriations committees, I expect that we could come up with a revised allocation plan and auction timetable."[58]

Public Safety Spectrum Needs

Efforts by government, industry, and public safety groups to replace outdated wireless communications systems for public safety agencies is also proving to be a challenge. The main obstacle in these efforts are (1) the high costs of new equipment, (2) the scarcity of unused spectrum, and (3) the need to coordinate among many organizations to enable public safety personnel to communicate with counterparts in other jurisdictions and government levels (known as "interoperability"). While progress has been made by public safety agencies, some argue that given the advances in technology, they should be closer to their goal. Some also compare the developments achieved by commercial wireless services to the status of public safety systems in terms of interoperability and ease of use, and argue that public safety systems should be further along.

Several legislative options could possibly expedite the development of a more unified public safety communications system. One possibility is to direct an increase in coordination among organizations working on public safety wireless communications issues, although introducing new bureaucratic requirements could cause delays. To provide spectrum for these systems, H.R. 4146 (Nick Smith, introduced March 30, 2000), contained a provision directing the FCC to allocate the spectrum between 139-140.5 MHZ, and between 141.5-143 MHZ, inclusive, to interoperability use by public safety services. There was no action on this bill. However, the 107th Congress could pursue legislation to make spectrum that has been reallocated for public safety available at a specified date, although incumbent television broadcasters using that spectrum would likely oppose such a measure. Additional spectrum for public safety may be found through sharing with DOD, as directed by the FY2001 Defense Authorization Act (P.L. 106-398). That law directs DOD, in consultation with the Departments of Justice and Commerce, to identify a portion of the 138-144 MHZ band to share in various geographic regions with public safety radio services.

---

[58] Letter from Michael Powell to Donald Evans, available at: [http://www.fcc.gov/Speeches/Powell/Statements/2001/stmkp127.pdf].

Regarding funding, for FY2001, $177 million has been provided for the Department of Justice to purchase new communications systems that will interoperate with state and local public safety systems. Other legislation to provide indirect funding for state and local wireless communications systems was introduced, but not enacted. Direct federal funding to state and local public safety agencies for new wireless communications systems would probably require significantly greater appropriations.[59]

In addition to efforts to enable public safety officials to communicate more effectively with each other, there is also an effort in government and industry to enable mobile telephone users to communicate with the public safety agencies in emergencies. This issue was addressed by the Wireless Communications and Public Safety Act of 1999 (P.L. 106-81, enacted October 26, 1999). While this law does not require any new spectrum allocations, it does require wireless service providers to deploy technologies that enable public safety officials to monitor the location of customers. Many are concerned that this requirement could violate the privacy rights of individuals, especially if third parties gain access to the location information of individuals. Some call for legislative or regulatory remedies to prohibit wireless service providers from releasing the location information.

Another public safety spectrum issue is interference in the 800 MHZ band between public safety radio communications and commercial mobile radio services. There has been an increasing number of reports of police and firefighter radios failing to function when used near commercial cell towers and base stations. While all providers are operating within the parameters of their FCC licenses, the FCC convened a working group of interested parties to discuss and study the issue. In February 2001, the working group and the FCC released a "Best Practices Guide" for avoiding interference between public safety wireless communications systems and commercial wireless communications systems for 800 MHZ. The FCC continues to collect information on this problem, and has stated that additional facts and analyses are needed to conclusively establish the causes of this interference and to identify potential remedies.

Spectrum Cap

To maximize competition in the wireless industry, in 1994 the FCC established rules to limit the amount of commercial mobile radio services (CMRS) spectrum any one entity can hold in any one market. Accordingly, no licensee in the cellular, PCS, or SMR services is allowed to have an attributable interest in more than 45 MHZ of CMRS spectrum within an urban geographic area. There is currently 180 MHZ licensed to these three services that is subject to the above "spectrum cap." In 1999, the FCC reviewed its spectrum cap policy, and decided that the spectrum cap continues to prevent any one entity from obtaining the majority of spectrum in a market, or warehousing spectrum for the purposes of shutting potential competitors out of a market.[60] The FCC ruled to maintain the cap with some modifications, such as allowing passive investors greater ability to fund small carriers. The ruling also raised

---

[59] For further discussion and analysis of this issue, see CRS Report RL30746, Wireless Communications Systems and Public Safety, November 27, 2000.

[60] FCC 99-, Docket No. 98-205, Report and Order In the Matter of 1998 Biennial Regulatory Review Spectrum Aggregation Limits for Wireless Telecommunications Carriers, released September 22, 1999.

the cap to 55 MHZ within rural geographic areas, and allowed for some waivers to be made to the spectrum cap (e.g., in the planned 700 MHZ band auction).

Many wireless service providers criticize the spectrum cap for preventing growth and innovation of their networks. They point to the higher spectrum caps in other nations (e.g., Japan and Britain), arguing that this has caused the U.S. wireless penetration rates to be lower than those of other industrialized countries. They further argue that even if the spectrum cap was justifiable in the past, it is no longer necessary given the strong competition for wireless services today. Some wireless service providers are reaching capacity with their current spectrum, leading to network congestion, busy signals, and delays for consumers. They argue that the best way to relieve congestion and continue their growth into new wireless services (such as 3G) is to repeal the spectrum cap.

In the 106th Congress, legislation was introduced November 16, 1999 (S. 1923), to prohibit the FCC from applying commercial mobile radio service spectrum aggregation limits (caps) to spectrum assigned by initial auction. This and a similar bill (H.R. 4758, introduced June 26, 2000) would have enabled incumbent mobile service operators to bid for 3G spectrum licenses. At a July 19, 2000, hearing of the House Commerce Committee, Subcommittee on Telecommunications, Trade and Consumer Protection, several witnesses from the wireless telecommunications industry recommended repealing the spectrum cap. The FCC, along with some wireless service providers, argued that the spectrum cap has produced the high level of competition that exists today and needs to be maintained. No action was ultimately taken on these bills in the 106th Congress.

In a recent report to Congress, the FCC decided to consider a notice of proposed rulemaking (NPRM) to be developed by its Wireless Telecommunications Bureau that will take "into consideration existing competitive conditions and technological developments that could affect the continued need for a cap."[61] On January 23, 2001, the FCC adopted an NPRM to reexamine the need for spectrum caps. The FCC is seeking comment on whether the spectrum caps for commercial mobile radio services should be eliminated, modified, or retained. Meanwhile, the Third-Generation Wireless Internet Act (S. 696), introduced by Senator Brownback on April 4, 2001, would require the FCC to lift spectrum aggregation limits (i.e., spectrum caps).

Spectrum Scarcity in Private Land Mobile Radio Services

A set of non-federal radio communications systems called Private Land Mobile Radio Services (PLMRS), including public safety, special emergency, industrial, land transportation, and radiolocation services, are treated differently from spectrum licensees whose main activity is providing commercial wireless communications services to the public.[62] Prior to the 1990s, spectrum allocated for PLMRS was divided into frequency bands representing 20 different types of communications service areas (see Table 2). For each PLMRS service area, the FCC designated a "frequency coordinator," to manage the PLMRS licenses in its service area. The FCC granted applications for new PLMRS licenses based on recommendations

---

[61] FCC 00-456, CC Docket No. 00-175, Report, In the Matter of the 2000 Biennial Regulatory Review, released
     January 17, 2001.
[62] PLMRS licenses are governed under 47 CFR, Part 90.

from the frequency coordinators in a given PLMRS service area. As a result of an increased number of requests from the wireless telecommunications industry for new PLMRS licenses, and the realization that insufficient spectrum was available to meet future needs, the FCC in 1992 began a proceeding to revise the policies governing PLMRS licenses. This proceeding is known as PLMRS "re-farming" (a reference to the redistribution of frequency allocations and rules for their use).

Table 2. Public Land Mobile Radio Service Areas/Frequency Pools

Public Safety Pool	Industrial/Business Pool
Local Government Radio Service	Power Radio Service
Police Radio Service	Petroleum Radio Service
Fire Radio Service	Forest Products Radio Service
Highway Maintenance Radio Service	Film and Video Production Radio Service
Forestry-Conservation Radio Service	Relay Press Radio Service
Emergency Medical Radio Service	Special Industrial Radio Service
Special Emergency Radio Service	Business Radio Service
	Manufacturers Radio Service
	Telephone Maintenance Radio Service
	Motor Carrier Radio Service
	Railroad Radio Service
	Taxicab Radio Service
	Automobile Emergency Radio Service

In 1995, as a first step, the FCC adopted a plan for PLMRS licenses below 800 MHZ in specific frequency bands and geographic areas.[63] These bands were previously divided into smaller frequency channels that were separated by unused spectrum. The 1995 ruling utilized the unused spectrum by creating new channels between the existing channels, and allowed licenses to be granted on the new channels for the use of new narrowband communications devices. The FCC determined that by using a new digital radio technology and by placing limits on the use of the new licenses, the amount of interference introduced on existing channels by the new licensees would be negligible.

Then in 1997, another FCC ruling consolidated the 20 PLMR services into two broad service pools – one designated as public safety and the other as industrial/business (see Table 2).[64] The frequency coordinators in the public safety pool continued to manage the same frequency bands as prior to consolidation.[65] The frequency coordinators in the industrial/business pool, however, were now able to accept applications from any service area

---

[63] The specific bands are: 150-174 MHz (nationwide), 421-430 MHz (only in Detroit, Buffalo, and Cleveland), 450-470 MHz (nationwide), and 470-512 MHz (shared with UHF-TV, available in 11 U.S. cities). FCC 95-255, PR Docket 92-235, Report and Order and Further Notice of Proposed Rule Making, "In the Matter of Replacement of Part 90 by Part 88 to Revise the PLMRS and Modify Policies Governing Them;" released June 23, 1995.

[64] FCC 97-61, Second Report and Order, released March 12, 1997, (effective October 1997). The ruling applied only to PLMRS licensees in spectrum bands below 800 MHz, including the 150-174, 421-430, 450-470, and 470-512 MHz bands.

[65] An exception to this provision was made to allow any frequency coordinator in the public safety pool to use frequencies allocated to Local Government Radio Service, which typically overlaps services provided by police, fire, and other public safety services.

within the pool if they determine that spectrum is available and would not cause harmful interference to incumbent licensees. An exception to that provision was made for railroad, power, and petroleum companies, which were deemed to provide critical public safety-related communications. Anyone else who wants a PLMRS license for the frequencies previously allocated exclusively to those three service groups must seek the recommendation of the frequency coordinators responsible for those services. New PLMRS licenses for frequencies that were previously shared among PLMRS licensees in the industrial/business pool, however, could be assigned by any of the frequency coordinators, and would be shared with existing licensees. The ruling contained several other provisions for the management of PLMRS. One important provision allowed "centralized trunking systems," which resell private wireless communications services using PLMRS spectrum on a leased basis, to operate the PLMRS licensees below 80 MHZ, with certain limitations to protect users sharing the same spectrum.[66]

In June 1998, two of the frequency coordinators submitted to the FCC an "Emergency Request for Limited Licensing Freeze." The request was submitted by the United Telecommunications Council (UTC, representing the power industry) and the American Petroleum Institute (API, representing the petroleum industry) to prevent new PLMRS licensees in the business/industrial pool to be granted on certain frequencies that were previously shared between the power and petroleum industries and other business/ industrial pool services. UTC and API argued that since the 1997 ruling went into effect, their industries had experienced an increased number of incidents of interference from other business/industrial pool licensees which posed a dangerous threat to power and petroleum operations. This request was followed by a Petition for Rulemaking, submitted by UTC, API, and the Association of American Railroads (AAR, represent the railroad industry). The petition asked the FCC to establish a new third pool for PLMRS licenses involved in public safety-related services, to protect those services from interference and encroachment by new industrial and commercial communications systems.

In April 1999, the FCC revised its 1997 ruling, requiring all PLMRS frequencies (both shared and exclusive) assigned to the power, petroleum, and railroad services prior to the adoption of the 1997 ruling to be coordinated by the frequency coordinators previously responsible for these services.[67] Applicants for these frequencies must obtain the concurrence of UTC, API, or AAR, as appropriate. The ruling also provided a similar protection to frequencies previously allocated to the former Automobile Emergency Radio Service, whose frequency coordinator is the Automobile Association of America.

In July 1999, two other frequency coordinators, the Manufacturers Radio Frequency Advisory Committee (MRFAC, representing the manufacturing industry) and Forest Industries Telecommunications (FIT, representing the forest products industry), petitioned the FCC to reconsider aspects of the 1999 ruling. The petition argued that the ruling gave a special status to frequencies allocated on a shared basis with the power, petroleum, and railroad radio services to the detriment of others in the industrial/business pool. The petition requested that the FCC stay the effective date of the ruling until this issue was addressed. Some argued that if auctions are implemented for some future PLMRS licenses, a radio

---

[66] A trunking system has the ability to automatically search all available radio frequencies for one that is not being used.

[67] FCC 99-68, Docket 92-235, Second Memorandum Opinion and Order, released April 13, 1999.

service that has a higher public safety status could possibly be exempted from an auction. Thus, creating a special pool for public safety-related services could create economic advantages for some licensees. In August 1999, the FCC granted a stay of its rules as requested by MRFAC and FIT until a final ruling on the matter was made.[68]

Legislation was introduced in the 106th Congress addressing the PLMRS issue (H.R. 866, introduced February 25, 1999) on which no action was taken. The bill would have created an advantage for the power or petroleum radio services over other Industrial/Business pool licensees by discontinuing licensing on frequencies formerly allocated to or near the power or petroleum radio services.

In November 2000, as part of its proceeding examining its auction authority, the FCC decided not to create a separate PLMRS license pool.[69] The FCC's Order also modified the PLMRS rules to allow licensees in the 800 MHZ Business and Industrial pool and the Land Transportation service area to convert their spectrum to commercial use under certain restrictions. The ruling also establishes that some PLMRS licenses (along with all private radio services) could be subject to auctions in the future if certain conditions (such as mutual exclusivity) apply. Although the FCC's ruling might have resolved the issues of managing PLMRS licenses, demand for PLMRS spectrum continues to be high. Disputes among the industries could again come to the attention of Congress if the parties are not satisfied with the ruling.

Attempts to Use Spectrum Auction Revenue for Other Programs

Current FCC policy, in accordance with current statue, is that potential revenue should not be used as the main argument for auctioning spectrum. Although many agree with that policy, revenue from spectrum auctions was used in the Balanced Budget Act of 1997 to help reduce the federal deficit. In addition, many bills were introduced in the 105th Congress to used spectrum auctions to offset various spending programs. None of those bills was enacted, and such attempts declined in the 106th Congress. This could be a result of decreased fiscal pressures due to the budget surpluses experienced during the past several years, as well as the acknowledgment that recent auctions have produced less revenue than the initial auctions.

However, at least one bill in the 106th Congress (S. 2762, introduced June 21, 2000) did contain a Sense of Congress statement advocating the use of spectrum auction revenue for a specific purpose. That bill, on which no action was taken, stated that "resources available through the auction of the analog [television] spectrum should be tapped to fund the development of a new educational and cultural infrastructure that utilizes today's technologies…" It is possible that legislation could be introduced again in the 107th Congress to use auction revenue for specific purposes other than reducing the federal debt, especially if federal budget pressures are exacerbated.

---

[68] FCC 99-203, PR Docket 92-235, Fourth Memorandum Opinion and Order, released August 5, 1999.
[69] FCC 00-403, WT Docket No. 99-87, Report and Order and Further NPRM, released November 20, 2000.

OTHER SPECTRUM-RELATED LEGISLATION IN THE 106TH CONGRESS

In addition to the legislation discussed in this report, several other bills were introduced in the 106th Congress that contained provisions concerning spectrum management. Some of them became law, while many did not.

- S. 376, introduced February 4, 1999 (the companion bill, H.R. 3261, was introduced in the House November 9, 1999) contained a provision prohibiting the FCC from assigning orbital locations or spectrum licenses to international or global communications services through the use of auctions. It further directed the President to oppose in the International Telecommunications Union and other international fora the use of auctions for such purposes. Enacted as the Open-market Reorganization for the Betterment of International Telecommunications Act (**P.L. 106-180**) on March 17, 2000.

- H.R. 783 (introduced February 23, 1999) sought to ensure the availability of spectrum to amateur radio operators by prohibiting reallocations of amateur radio service and amateur satellite service spectrum unless the FCC provides equivalent replacement spectrum. Referred to House Commerce Committee, Subcommittee on Telecommunications, Trade and Consumer Protection. Companion bill S. 2183 (introduced March 6, 2000) referred to Senate Committee on Commerce, Science and Transportation. No further action.

- H.R. 879 (introduced February 24, 1999) would have exempted licenses in the instructional television fixed service from auctions. Referred to House Committee on Commerce, Subcommittee on Telecommunications, Trade and Consumer Protection. No further action.

- H.R. 1554 (introduced April 26, 1999) contained a provision (Sec. 5008, "Community Broadcasters Protection Act") and companion bill S. 1547 (introduced August 5, 1999) seeks to preserve low-power television (LPTV) stations that provide community broadcasting by directing the FCC to grant qualifying LPTV stations special "class A" licenses and find alternative spectrum on which to locate those stations wherever full power stations are given priority to their spectrum in the transition to DTV. Incorporated into the Intellectual Property and Communications Omnibus Reform Act of 1999 (S. 1948, Section 5008, which was incorporated by cross-reference in the conference report H.Rept. 106-479 to H.R. 3194, FY2000 Consolidated Appropriations bill) and enacted (**P.L. 106-113**) on November 29, 1999.

- H.R. 2379 (introduced June 29, 1999) sought to ensure that biomedical telemetry operations are provided with adequate spectrum to support that industry's existing and future needs, and requires a report from the FCC on future spectrum needs of telemedicine and telehealth providers. Referred to House Committee on Commerce, Subcommittee on Telecommunications, Trade and Consumer Protection. No further action.

- H.R. 2630 (introduced July 29, 1999) NTIA Reauthorization Act, contained a provision directing NTIA to conduct an assessment of spectrum reallocations to non-federal use and the implications of such reallocations for affected federal agencies. Although the bill was not enacted, a similar provision was added to the FY2000 Defense Authorization Act (**P.L. 106-65**).

- S. 1824 (introduced October 28, 1999) would have directed the FCC to: (1) identify and allocate at least 12 MHZ of spectrum located between 150-200 MHZ for use by private wireless licensees on a shared-use basis; (2) reserve at least 50 percent of the reallocated spectrum for private wireless systems; and (3) reallocate and assign licenses for such spectrum. It further required the FCC to devise a schedule for payments to the Treasury for shared spectrum used by private wireless systems, and adopt a payment schedule in the public interest. Referred to House Committee on Commerce. No further action.

- H.R. 3615 (introduced March 10, 2000) would have provided loan guarantees to rural television stations for improvements related to producing local broadcasts in underserved areas, but prohibited the use of the funds for spectrum auctions. Reported by the House Committee on Agriculture, Commerce, and Rules. Passed the House (amended) April 13, 2000. No further action.

- S. 2454 (introduced April 13, 2000) authorizes low power television (LPTV) stations to provide "data-casting" (e.g., financial or Internet-related) services to subscribers. Referred to Senate Committee on Commerce. A similar provision was inserted in the FY2001 Omnibus Consolidated and Emergency Supplemental Appropriations Act (H.R. 4577), enacted December 21, 2000 (**P.L. 106-554**).

- H.R. 4758 (introduced June 26, 2000) would have prohibited the FCC from applying limits on spectrum aggregation to any license for commercial mobile radio service granted in an auction held after January 1, 2000, or to any subsequent application for the transfer or assignment of such a license. Referred to House Committee on Commerce, Subcommittee on Telecommunications, Trade and Consumer Protection. No further action.

SPECTRUM-RELATED LEGISLATION IN THE 107TH CONGRESS

H.R. 817 (Bilrakis)

Amateur Radio Spectrum Protection Act of 2001. Prevents the FCC from reallocating or diminishing frequencies used by amateur radio and amateur satellite services. Introduced March 1, 2001; referred to Committee on Energy and Commerce.

S. 549 (Crapo)

Amateur Radio Spectrum Protection Act of 2001. Prevents the FCC from reallocating or diminishing frequencies used by amateur radio and amateur satellite services. Introduced March 15, 2001; referred to Committee on Commerce, Science, and Transportation.

S. 696 (Brownback)

Third-Generation Wireless Internet Act. Prohibits the Federal Communications Commission from applying spectrum aggregation limits to spectrum assigned by auction after December 31, 2000. Introduced April 4, 2001; referred to Committee on Commerce, Science, and Transportation.

CONCLUDING REMARKS

The growth in demand for wireless services has been unprecedented, with estimates that, by 2002, wireless users will number up to one billion worldwide.[70] Whether the U.S. wireless industry will keep up with or surpass the growth and penetration rates of other industrialized nations, depends in part on how the radio spectrum is managed both here and abroad. The wireless services industry is highly dynamic, spurred by the growth of electronic commerce and development of wireless Internet access services. Federal, state, and local government spectrum needs are also increasing as those services strive to maintain the same level of technology as the private sector.

The pace of the wireless revolution demands that new spectrum resources be found quickly. The challenge is to manage the spectrum resources to maximize the efficiency and effectiveness of meeting these demands without disenfranchising incumbent spectrum users, while ensuring that competition is maximized and that consumer, industry, and government groups are treated fairly. Congress will likely play a significant role in ensuring that those challenges are met.

---

[70] Testimony of Tom Sugrue, Chief, Wireless Telecommunications Bureau, FCC, before the House Commerce Committee, Subcommittee on Telecommunications, Trade and Consumer Protection, July 19, 2000.

APPENDIX 1. SPECTRUM LICENSES AUCTIONED TO DATE

License type and geographic area	Amount of spectrum per license and its uses	Number of licenses sold	Date closed	Net High bids, $millions
Narrowband: nationwide	50 kHz (paired), 50/12.5 kHz, and 50 kHz: paging, messaging	10	7/94	617.0
IVDS: MSAs	0.5 MHZ: interactive data	594	7/94	213.9
Narrowband PCS: regional	50 kHz (paired), 50/12.5 kHz: paging messaging	30	11/94	392.7
Broadband PCS A&B blocks: MTAs	30 MHZ: mobile voice and data	99	3/95	7,019.4
DBS: nationwide (two separate auctions)	Shared spectrum* Subscription television service	2	1/96	734.8
MDS: BTAs	Heavily encumbered spectrum* Subscription TV broadcast	493	3/96	216.2
900 MHZ SMR: MTAs	25 kHz: mobile dispatching	1,020	4/96	204.3
Broadband C-block PCS: BTAs	30 MHZ: mobile voice and data	493	5/96	9,197.5
Broadband C-block PCS re-auction: BTAs	30 MHZ: mobile voice and data	18	7/96	904.6
Broadband PCS blocks D, E, and F: BTAs	10 MHZ: mobile voice and data	1,472	1/97	2,517.4
Unserved cellular areas: MSAs, RSAs	25 MHZ (encumbered): mobile voice and data	14	1/97	1.8
Wireless communications service: Major/Regional Economic Areas	10 MHZ and 5 MHZ: multiple wireless uses	126	4/97	13.6
Digital audio radio services: nationwide	12.5 MHZ: satellite radio broadcasting	2	4/97	173.2
Upper 800 MHZ SMR: Economic Areas	1 MHZ, 3 MHZ, and 6 MHZ: mobile voice and data	524	12/97	96.2
LMDS: BTAs	1,150 MHZ and 150 MHZ: fixed voice, data, and video	864	3/98	578.7
220 MHZ band: economic area/grouping, nationwide (two separate auctions)	100 kHz and 150 kHz (encumbered): voice, data, paging, fixed services	915	10/98 6/99	23.6
VHF Public Coast: Public Coast station areas	25 kHz (paired): maritime/ship fixed/mobile communications	26	12/98	7.5
C, D, E, and F Block Broadband PCS: BTAs	30 MHZ, 15 MHZ, and 5 MHZ (paired): mobile telephone service	302	4/99	412.8

License type and geographic area	Amount of spectrum per license and its uses	Number of licenses sold	Date closed	Net High bids, $millions
Locations and Monitoring Service: Economic Areas	6 MHZ, 2.25 MHZ, and 5.75 MHZ with 250 kHz (paired): radio signals to determine location/use of mobile phones	289	3/99	3.4
LMDS re-auction: BTAs	1,150 MHZ and 150 MHZ: multiple wireless uses	161	5/99	45.1
"Closed" broadcast station construction permits: no geographic limit defined	10 kHz (AM), 200 kHz (FM), 6 MHZ (television, LPTV): radio and TV broadcasting	115	10/99	57.8
929-931 MHZ paging services: major economic areas	20 kHz: paging/messaging services, data transmission	985	3/00	4.1
Broadcast construction permits (three separate auctions)	200 kHz (FM), 6 MHZ (television, LPTV): radio and TV broadcasting	4	10/99 3/00 7/00	20.2
39 GHz band licenses: economic areas	50 MHZ (paired): fixed/mobile two-way communications	2,173	5/00	410.6
700 MHZ guard band: major economic areas	2 MHZ (paired) and 1 MHZ (paired): leased wireless services subject to strict interference limits	96	9/00	519.9
800 MHZ SMR (general category): economic areas	1.25 MHZ (split into six channels): voice/data/paging and other wireless services	1,030	9/00	319.5
Reclaimed C&F block PCS licenses, 1890-1975 MHZ	Two 15 MHZ (paired) or three 10 MHZ (paired) (C-block, 10 MHZ (paired) F-block; mobile voice and data	2,800	12/00	29.0
700 MHZ guard band: unsold licenses from previous 700 MHZ guard band auctions	2 MHZ (paired) and 1 MHZ (paired): leased wireless services subject to strict interference limits	422	1/01	16,857.0
VHF Public Coast (156-162 MHZ) and LMF licenses (902-928 MHZ)	6 MHZ Public Coast (marine communications); 6, 2.25, 5.75 MHZ LMF (radiolocation)	8	2/01	21.0
Total				$41,614.1 M

Source: CRS, based on data from the FCC

* Due to the DBS rules for spectrum channelization and the existence of many licensees encumbering the MDS spectrum, the amount of spectrum for the DBS and MDS auctions cannot be easily established.

APPENDIX 2. ONGOING AND FUTURE SPECTRUM AUCTIONS

License Type: Geographic Area	Frequency Range	Auction Date
700 MHZ Band	746-764 MHZ and 776-794 MHZ	September 12, 2001
Lower 700 MHZ band	698-746 MHZ	Not yet scheduled
FM Broadcast	88-108 MHZ	December 5, 2001
AM Filing Window	0.535-1.065 MHZ	Not yet scheduled
24 GHz (DEMS)	Around 24 GHz	Not yet scheduled
4.9 GHz	4940-4990 MHZ	Not yet scheduled
Narrowband PCS: MTA and BTA	50 KHz paired licenses in the 900 MHZ range	Not yet scheduled
LPTV	512-806 MHZ	Not yet scheduled
Paging (lower channels	Licenses of varying sizes between 35-930 MHZ	June 26, 2001
218-219 MHZ (formerly IVDS): MSA and RSA	500 kHz licenses at 218 and 219 MHZ	Not yet scheduled
Public Coast and Location and Monitoring Services	156-162 MHZ and 904-928 MHZ	June 6, 2001
GWCS	4600-4685 MHZ	Not yet scheduled

Source: CRS, based on data from the FCC.

Chapter 2

PUBLIC BROADCASTING: FREQUENTLY ASKED QUESTIONS

Bernevia McCalip

SUMMARY

Public broadcasting was inaugurated by Congress in 1967 with the statutory chartering of the Corporation for Public Broadcasting (CPB), a private, nonprofit corporation mandated to provide nationwide programming to further the educational and cultural interests of the American people. Such programming was to constitute an expression of "diversity and excellence." The CPB provides grants to qualified public radio and television stations for use at their discretion for purposes related to program production or acquisition and general operations. The corporation also supports the production and acquisition of broadcast programming for national distribution; assists in the financing of several systemwide activities, such as national satellite interconnection services; and provides limited technical assistance, research, and planning services to improve systemwide capacity and performance. The CPB is accountable to Congress for its faithful execution of its statutory mandate, expenditure of appropriated funds, and adherence to policy conditions attending the provision of such monies.

WHAT IS PUBLIC BROADCASTING?

Public broadcasting is a community of local radio and television broadcast stations. Nationwide, there are approximately 1,045 such stations. Most public broadcasting stations are run by universities, nonprofit community associations, and state government agencies. Regarded as fundamentally a public service, the public broadcasting system observes the principle of local autonomy. That is, public broadcasters are free to make programming decisions independent of Congress and the Corporation for Public Broadcasting (CPB) as to what will be available to their listening or viewing audience, as well as on their programming schedule.

WHAT IS THE CORPORATION FOR PUBLIC BROADCASTING?

The CPB is a private, non-profit corporation responsible to Congress for the success of public broadcasting in the United States. The CPB is guided by a nine-member board of directors (including its president) appointed by the President with the advice and consent of the Senate. The members serve for staggered six-year terms. The CPB was established as a private, nonprofit corporation to eliminate political influence. By law, the CPB is required to make reports to Congress and submit to audits.

WHAT IS THE RELATIONSHIP OF THE CPB TO THE FEDERAL GOVERNMENT?

The CPB is accountable to Congress for the performance of its public broadcasting responsibilities. However, it is an entirely independent entity and is not a government agency. The CPB allocates federal funds referred to as Community Service Grants (CSGs), to public radio and television stations nationwide and is accountable to Congress on the use of such monies.

WHAT IS CPB'S RELATIONSHIP WITH LOCAL BROADCASTING STATIONS?

The CPB provides financial support and a variety of services to public radio and television stations nationwide to ensure universal access to public broadcasting's educational services and programs. The CPB ensures that the stations can exchange program materials through a national system of interconnection and that stations serve their communities effectively.

WHAT IS THE FUNCTION OF THE CPB?

The CPB's principal function is to receive and distribute government contributions (or federal appropriations) to fund national programs and to support qualified member radio and television stations based on legislatively mandated formula; the bulk of these funds are to provide CSGs to member stations that have matching funds. The CPB also promotes public telecommunications serves (television, radio, and online) for the American people.

WHAT ARE THE FUNDING SOURCES FOR PUBLIC BROADCASTING?

The public broadcasting system receives funding from many sources. These sources include members (or subscribers); state governments; businesses; federal and local governments; state, private, and public colleges; foundations; federal grants and contracts; and auction receipts. Members (or subscribers) are the largest single source of funding for public radio and television programming.

HOW IS THE CPB FUNDED?

The operation of the CPB is financially supported by Congress through annual appropriations. The CPB received its first funding in FY 1969 and continued to receive annual appropriations until FY1975, when the Public Broadcasting Financing Act of 1975 (P.L. 94-192) establishes authorization for advanced or long-term financing. The 1975 Act authorized funding for over a five-year period. Advanced funding from Congress now sets actual appropriations two years in advance of stations receiving their funds from CPB.

WHAT METHOD IS USED TO ALLOCATE FUNDS TO LOCAL BROADCASTING STATIONS?

The CPB's appropriations are allocated through a distribution formula established in its authorizing legislation. By law, 6% is allocated for system support and 5% for CPB operations, and the remaining 89% is distributed in the form of CSGs to radio (25%) and television (75%) stations. The CSGs help public broadcasters produce and acquire programming, finance production equipment and facilities, launch community outreach services in connection with public service programming, and pay for satellite interconnection services.

HOW ARE CSG'S DETERMINED?

The CSG is divided into two parts: a base grant, which is a fixed amount, and an incentive grant designed to reward a station according to the amount of the federal appropriation and the non-federal funds that a station raises. Every station qualifies for the incentive grant. Only one base grant is available in a market. In markets with more than one qualifying public television, each station may receive a partial base grant that, when combined, equals a single base grant. For FY2000, the base grant per station totaled $330,000.

HOW IS THE PUBLIC BROADCASTING SYSTEM STRUCTURED?

At the apex of the public broadcasting system is the CPB. The Public Broadcasting Service (PBS) was created by the CPB in 1969 to operate and manage a national interconnection (or satellite) system and to provide a distribution channel for national programs to public television stations. National Public Radio (NPR) was created in 1970 as a news-gathering, production, and program-distribution company. Both the PBS and NPR are owned, supported, and operated by member stations.

DOES THE CPB HAVE ANY CONTROL OVER THE SELECTION OF PROGRAMS AIRED BY PUBLIC RADIO AND TELEVISION STATIONS?

The Public Broadcasting Act of 1967 requires the CPB to "afford maximum protection from extraneous interference and control" in the development and distribution of programming. However, the same provision directs the CPB to provide protection against interference in local programming decisions. This dos not mean that local stations may operate free of any concern for responsible editorial judgments. It does mean that both Congress and the CPB regard the local community as the best evaluator of programming that meets the community's needs and interests.

ARE PUBLIC BROADCASTERS REQUIRED TO PROVIDE PUBLIC INTEREST AIR TIME TO POLITICAL CANDIDATES?

The Communications Act of 1934, as amended (47 U.S. Code 399), prohibits the public broadcasting system from engaging in any editorializing, supporting, or opposing of political candidates.

WHO DETERMINES WHAT PROGRAMS ARE APPROPRIATE FOR PUBLIC TELEVISION?

At the direction of Congress, public broadcasting supports production proposals that generate constructive, respectful discussion about controversial topics. The CPB's mandate is to *facilitate* the development of programs of high quality, diversity, creativity, excellence, and innovation, which are obtained from diverse sources, with strict adherence to objectivity and balance in all programs or series of programs of a controversial nature. The CPB's role in program development is, by law, limited. It does not review or edit the contents of the programming it funds. The CPB, by law, has no say in the production of many of the programs it funds. And, the CPB, by law, has no say in individual radio and television stations' programming decision. Each individual radio and television station uses independent discretion as to the type and kind of programs that best meet the needs of its listening audience.

HOW DOES DIGITAL BROADCASTING
AFFECT PUBLIC BROADCASTING?

Broadcasting by digital signal,[1] beginning on May 1, 2002, is a Federal Communications Commission requirement. This same federal mandate requires that television broadcasters deliver both an analog and a digital signal by 2003. Current estimates are the cost of developing high-definition television programming, for broadcasting by digital signal, cost about 15% to 30% higher than the cost of current (analog) program production methods. The total estimated cost of digital conversion is $1.8 billion.

The arrival of digital broadcasting, however, promises to enable public broadcasting stations to expand their services. With the capability to multicast many channels, even the smallest public television stations will be able to transmit data-enhanced digital services to the hearing – and visually impaired and multilingual populations. Special programs, such as job training, adult education, and literacy training, are additional services public broadcasting will be able to offer under the digital signal.

---

[1] For further information about this type of broadcasting, see CRS Report 97-925, *Digital Television: Recent Developments and Congressional Issues*, by Richard Nunno.

Chapter 3

TELEPHONE BILLS: CHARGES ON LOCAL TELEPHONE BILLS

James R. Riehl

TELEPHONE COMPANIES

According to estimates of the Federal Communications Commission (FCC), there are over 1,300 companies that provide local telephone services and over 700 companies that provide long-distance telephone services in the United States.[1] There may be almost that many ways of presenting a telephone bill to a customer. The FCC does not dictate the form or wording of a telephone bill. State public utility commissions, the entities that oversee telephone industry regulation within each state, generally do not try to control form and wording of telephone bills either.

COALITION FOR AFFORDABLE LOCAL AND LONG-DISTANCE SERVICES (CALLS)

In August 1999, six of the largest phone companies (AT&T, Sprint, Bell Atlantic, BellSouth, SBC, and GTE) announced an industry plan to substantially revise the complicated system of telephone access charges,[2] which include the subscriber line charge (SLC, see the section Subscriber Line Charge) and the presubscribed interexchange carrier charge (PICC, see the section Presubscribed Interexchange Carrier Charge). The plan, referred to as CALLS, was modified by the coalition of phone companies after criticism from the FCC and consumer groups. The FCC adopted the main provisions of this 5-year access

---

[1] U.S. Federal Communications Commission. *Statistics of Communications Common Carriers.* 2000/2001ed. Washington, 2001. p. v-vi.
[2] Access charges are the fees that long-distance companies pay to local telephone companies for access to the local phone network.

reform plan on an interim mandatory basis on May 31, 2000.[3] The access charge rate structure is mandatory for all major local phone companies with certain rate level components being mandatory on an interim basis. The mandatory nature of the plan has been criticized by some companies who do not believe that the plan does enough to guarantee affordable local telephone service in rural, high-cost areas. The plan permits some companies to opt out after the first year. Those who opt out will be subject to special cost studies. Some major long-distance and local telephone companies that were not parties to the proposal and certain consumer groups have criticized the plan. US West, recently acquired by Qwest and the only large local phone company that did not agree to the CALLS plan, and the National Association of State Utility Consumer Advocates (NASUCA) both filed petitions in court for review of aspects of the CALLS plan. US West believed the plan was "arbitrary, capricious and otherwise contrary to law." NASUCA representatives stated that the plan would actually raise phone bills. On September 19, 2000, Qwest altered its position concerning CALLS and agreed to carry out its provisions. Qwest also announced that it would drop various lawsuits that it had filed and would review others.

ACCESS CHARGES

Subscriber Line Charge (SLC)

The subscriber line charge is a federally regulated charge that first appeared on phone bills following the divestiture of the American Telephone & Telegraph Company (AT&T) in 1984. It is also referred to as an "access charge" and is intended to allow local telephone companies to recover some of the fixed costs (telephone wires, poles, and other facilities) of connecting phone customers to the interstate long-distance network. When a customer makes an interstate long-distance call, in the vast majority of cases he/she must use a local phone company's network to connect to the long-distance network. Access charges are paid to local telephone companies by both the end user (business or residential customers) and the long-distance company carrying a long-distance call. The SLC paid by end users appears on a bill as a specific itemized charge. The long-distance company that carries an individual long-distance call pays access charges to both the local phone company originating the call and the one terminating the call. The access charges paid by the long-distance carriers do not appear on a telephone bill. Over the past few years, the FCC reduced the amount of access charges paid by long-distance companies. Access charges are kept by the local phone companies. They are not forwarded to the federal government. An FCC Consumer Fact Sheet on the SLC may be obtained at [http://www.fcc. gov/cib/consumerfacts/SLC061500.html].

In conjunction with decisions related to the implementation of the Telecommunications Act of 1996. The FCC revised the SLC for residential and business customers with more than one telephone line, although SLC charges for customers with a single line did not change. In most cases, until the CALLS revisions, the SLC for a primary residential line was $3.50 per month. Any additional residential lines are considered non-primary lines. The SLC for non-primary lines was capped at $5 per line per month through 1998. Starting in 1999, the SLC

---

[3] U.S. Federal Communications Commission. *In the Matter of Access Charge Reform...*, FCC 00-193. Adopted May 31, 2000. Released May 31, 2000. Available via the FCC Web site at [http://www.fcc.gov/ccb].

for non-primary residential lines was adjusted for inflation and increased $1. It was capped at $6.07. However, this did not mean that all non-primary lines incurred a $6.07 charge on a telephone bill. If the local telephone company's average interstate costs of providing that line were less than $6.07, it could only charge the actual amount of its costs to a consumer.

CALLS **Revisions** *of the SLC.* As of July 1,2000, the SLC and PICC for **residential and single line businesses** were consolidated into a new SLC. The PICC charge was eliminated as a separate charge. The new primary line residential and single-line business SLC was capped at **$4.35** per month and on July 1, 2001 rose to $5. Under this plan, the cap will rise to $6 on July 1,2002, and to $6.50 on July 1, 2003. The increases after 2001 are subject to FCC validation. The FCC notes that for the first year, the new single SLC charge was lower than the separate SLC and PICC charges combined.

The SLC cap for residential customers and single-line businesses remained at $3.50 for smaller local telephone companies (approximately 1,300 carriers providing service to less than 10% of total telephone access lines). On October 11, 2001, the FCC adopted an order to reform these charges also.[4] In this order, referred to as the MAG (Multi-Association Group) Plan, the SEC caps for the smaller carriers were increased to the same levels paid by most other telephone subscribers. As of January 1, 2002, the SEC cap for residential and single-line businesses will increase to $5. Beyond that date, the FCC will conduct cost review studies, and the cap may increase to $6 on July 1, 2002, and to $6.50 on July 1, 2003.

Under the CALLS plan, **non-primary line residential** (two or more lines in the home) SLC charges were increased and capped at $7 beginning July 1, 2000. The charge will remain at this level for 5 years. However, if the telephone company s average interstate costs of providing the line are less than $7, it may only charge the customer the amount of its costs. Not all non-primary residential lines will be charged at the $7 cap. Prior to the implementation of CALLS, this charge was capped at $6.07, but was scheduled to increase by $1 plus an amount for inflation on July 1, 2000.

In the MAG Plan, the FCC adopted the MAG proposal to apply the same SLC caps to primary **and** non-primary residential lines of the smaller carriers. The FCC stated that several commenters in this proceeding indicated that higher SLC rates for non-primary lines would limit the growth of these lines, which are often used for advanced telecommunications services and are an important source of revenue for the smaller carriers.

Prior to CALLS, the maximum SLC for **businesses with multiple** lines was $9 per line per month through 1998. In 1999, the multiple line business SLC was adjusted for inflation and increased to $9.20 per line. This charge was adjusted for inflation annually. As with the residential SEC. local phone companies could only recover their costs. Thus, business customers with multiple lines did not necessarily see a $9.20 charge for each line. The amount could be less, and according to the FCC, the average SLC for businesses with multiple lines was $7.17. As a result of the adoption of the MAG Plan order, the multiple line SLC cap for smaller carriers will increase to $9.20 on January 1, 2002.

Multiple line businesses will **NOT** see a consolidation of the PICC and SEC charges. Under the CALLS plan, the **multiple line business PICC** is capped at $4.31 (its pre-CALLS cap) and will be reduced and eliminated in most areas over the next several years (or sooner).

---

[4] U.S. Federal Communications Commission. In the Matter of Multi-Association Group Plan for Regulation of Interstate Services of Non-Price Cap Incumbent Local Exchange Carriers and Interexchange Carriers, FCC 01-304. Adopted October 11, 2001. Released November 8, 2001. Available via the FCC Web site at [http://Iwww.fcc.gov/ccb].

The **multiple line business *SLC* will be frozen for** 5 years. For business customers of the largest local phone companies, the SLC charge is $9.20 or the company's average interstate cost of providing the line in that state, whichever is less. Multiple line business customers of the smaller local telephone companies will be charged $6 or the cost of providing the line in that state, whichever is less. The FCC plans to reevaluate the multiple line business charges at the end of the 5-year period covered by the CALLS plan.

The presence of a cap does not mean that every customer will be charged that specific amount on their bill. The cap is the maximum charge that may appear. The actual charge on an individual phone bill may be lower than the cap.

Presubscribed Interexchange Carrier Charge (PICC)

The **PICC** began appearing on telephone bills in January 1998. It was a flat-rate per-line charge that long-distance companies paid to local telephone companies. It was charged in addition to the SLC, because the FCC determined that the SLC did not allow local phone companies to recover all of the fixed costs associated with the interstate portion of the local loop. The FCC set PICC charges as ceilings, not absolute rates, and thus specific PICCs varied from state to state depending upon the costs of providing service within the state. The charge could be assessed for all telephone lines regardless of whether a business or residential customer had actually selected (presubscribed) a specific long-distance company.

As of July 1, 1999, the PICC for primary residential lines and businesses with a single line was capped at $1.04 per month, up from $0.53 in 1998. The primary line and single line business PICC was adjusted annually for inflation and increased by $0.50. Through June 30, 2000, the maximum PICC charge for non-primary residential lines was $2.53 per line per month, up from $1.50 in 1998. The cap for business customers with multiple phone lines was raised to $4.31 per line per month, up from $2.75 in 1998. The multiple business line PICC ceiling could be adjusted for inflation and increased, if necessary, by approximately $1.50 per year. As with the residential and single-line business PICC, the FCC estimated that, as its plans were implemented. PICC charges would decrease and eventually reach zero in many places.

Long-distance companies took various approaches to including or not including PICC charges on phone bills. In some cases, the charges appeared as an itemized line on a bill, but they also may have been lumped in with other charges and labeled "national access fee" or "carrier line charge." The FCC did not order long-distance companies to present PICC charges in a specific way, nor did the FCC order the companies to charge the customer directly for PICC charges. The FCC stated that its reductions in access charges which the long-distance companies pay to local phone companies largely offset any increases in per-line or other charges, making them revenue-neutral. Some long-distance companies chose to recover all or part of the PICC charges from their customers and stated that they had to do so because their costs rose and the FCC reductions in access charges were not enough and had already been passed on to customers. Long-distance companies requested further reductions of these charges.

CALLS Revisions of the PICC

Under the CALLS plan, on **July 1, 2000,** the PICC charge was eliminated as a separate charge for **residential and single line** business customers. The PICC and SLC (see section Subscriber Line Charge) were consolidated into a single, new SLC charge. However, multiple line businesses will NOT see a consolidation of the PICC and SLC charges. The multiple line business **PICC** will be capped at $4.31 (its pre-CALLS cap) and will be reduced and eliminated inmost areas over the next several years (or sooner). As with SLC caps, the presence of a PICC cap does not mean that all customers will be charged at the cap rate. The specific charge may be less than the cap and must be based on the actual cost of providing phone service in each area. An FCC Consumer Fact Sheet on the PICC charge is available at [http://www.fcc.gov/cib/consumerfacts/PICCchanges.html].

Other CALLS Provisions

Overall, according to the FCC, this action will simplify charges and reduce the fees appearing on monthly bills, especially for low-volume residential and business users. Due to the wide variety of billing formats, the different fees on telephone bills, charges attached to different calling plans, and the volume of calls a customer makes, it is not possible to state that any particular bill will decrease by a specific amount or percent. The FCC has stated, however, that low-volume users (30 minutes or less of long-distance calling per month) may save between $10 and $50 per year. Various observers believe it is even more difficult to quantify savings for heavy users of long-distance services. In addition, this action by the FCC may result in an increase in local and long-distance competition and further reductions in long-distance charges.

According to the FCC, some of the major consumer benefits of the CALLS plan are:

- the elimination of the residential and single line business PICC. The multiple line business PCC will be reduced over time and eliminated in some areas.
- a $3.2 billion reduction in access charges paid by long-distance companies to local phone companies.

Although major long-distance companies have agreed to pass these savings on to their customers over the 5-year life of the CALLS plan, some consumer groups and analysts question whether all of the savings would actually reach consumers, and when. The government cannot force the companies to pass on the savings. It is not clear at this time whether per minute charges for long-distance calling will fall.

- availability of at least one long-distance plan (to AT&T and Sprint customers) that does not have a monthly minimum use charge.

Monthly minimum use fees, (see the section "Minimum Use Fees") of approximately $3 per month have been charged by some long-distance carriers to customers who do not make large volumes of long-distance calls. Although this plan does not abolish the use of minimum monthly fees and the FCC did not reach the conclusion that such fiat fees were unreasonable, inequitable, or inconsistent with the Communications Act, companies agreed to eliminate or make avoidable some of these fees.

As part of the CALLS proposal, AT&T and Sprint have agreed to make long-distance plans available that would address the needs of low-volume users. Also, members of the coalition agreed to work with the Consumer Information Bureau at the FCC to develop a consumer education plan. The plan will address important issues relating to long-distance and local phone service pricing and service. In addition, the CALLS companies will create programs and materials to assist consumers in understanding their telephone bills. In order to reach the maximum number of consumers, the materials must be available in various formats and languages. Within 90 days of publication of provisions of the Order in the *Federal Register,* CALLS companies had to submit a compliance statement relating to their consumer education plan to the FCC.[5] Education efforts must continue over the 5-year life of the plan. A report was filed with the FCC on September 19, 2000.

As part of the education effort, the CALLS member companies established Web sites to provide phone bill assistance and other information to consumers. The Web sites are [http://www.phonebillcentral.org] and [http://www.lifelinesupport.org].

- identification of $650 million in implicit universal service support and establishment of an explicit universal service support mechanism to replace the implicit support

According to the FCC, phone companies were collecting approximately $650 million in universal service (see the section "Universal Service") support for high-cost customers through their access charges. Under the new rules, this money is removed from access charges and replaced with an assessment on all telecommunications carriers' interstate revenues. The money will be placed in a new universal service mechanism (separate from the existing high-cost fund) and made available to any carrier serving customers in high-cost areas. This new mechanism is capped at $650 million and is targeted to density zones and study areas that have the greatest need for it. As of July 1, 2000, price cap local phone companies must create a separate line item to recover all contributions to the universal service support mechanism.

For more information on the CALLS plan and its provisions, consult the FCC Web site at [http://www.fcc.gov/cib/consumerfacts/Calls2.html] or contact them at:

Federal Communications Commission
Consumer Information Bureau
445 12th Street, S.W.
Washington, D.C. 20554
1-888-CALL-FCC

---

[5] *Federal Register*, June 21, 2000, p. 3684-704.

TRUTH-IN-BILLING AND BILLING FORMAT

On September 17, 1998, the FCC adopted a Notice of Proposed Rulemaking addressing the issue of the clarity of telephone bills.[6] The three main proposals of the rulemaking were:

- Telephone bills should be clearly organized and highlight any new charges or changes to consumers' services;
- Telephone bills should contain full and non-misleading descriptions of all charges and clear identification of the service provider responsible for each charge; and
- Telephone bills should contain clear and conspicuous disclosure of any information consumers need to make inquiries about charges.

The FCC received over 60,000 consumer inquiries concerning telephone bills in 1998. On April 15, 1999, the FCC issued an Order generally adopting the proposed principles and minimal, basic guidelines to help consumers understand their telephone bills.[7] The guidelines adopted implement three basic principles. Consumers should know:

- who is asking them to pay for service;
- what services they are being asked to pay for; and
- where they can call to obtain additional information about the charges appearing on their telephone bill.

The FCC chose to adopt broad, binding principles instead of detailed rules that would rigidly control all of the wording and the format of a telephone bill. Thus, telephone companies have wide latitude to satisfy the adopted principles in a way that serves the needs of the carrier and the customer. In its Order, the FCC states that:

"We incorporate these principles and guidelines into the Commission's rules, because we intend for these obligations to be enforceable to the same degree as other rules. Thus, while we provide carriers flexibility in their compliance, we fully expect them to meet their obligation to provide customers with the accurate and meaningful information contemplated by these principles."

Deniable and Non-deniable Charges

The Truth-in-Billing Order also requires companies to identify charges on a customer's bill that are "deniable" and "non-deniable." Generally, deniable charges are those that, if not paid, may result in the termination (denial) of a customer's local telephone service. Non-deniable charges are those that, if not paid, will not result in termination of the customer's local telephone service. No specific format on a bill is required, although deniable and non-deniable charges must be clearly and conspicuously identified. In addition, carriers are free to

---

[6] *Federal Register,* October 14, 1998, p. 55077-83.
[7] U.S. Federal Communications Commission. *In the Matter of Truth-in-Billing and Billing Format,* First Report and Order and Further Notice of Proposed Rulemaking, CC Docket 98-170, FCC 99-72. Adopted April 15, 1999. Released May 11. 1999. Available via the FCC Web site at [http://www.fcc.gov/ccb].

choose other methods of informing consumers about charges that may be contested. State laws may also address this issue.

The FCC views identification of charges into these two categories as protecting consumers from paying questionable, unauthorized charges out of fear of having their local telephone service disconnected. However, this guideline applies only to companies who include both categories of charges on a single bill. Companies that bill directly for a service that includes no basic local telephone service would not be covered. For example, customers being billed directly by a wireless telephone company for only wireless service would not have charges for wired basic local telephone service on their bills. Although not paying charges on the wireless bill would have no effect on their at-home wired service, non-payment of the wireless bill may result in termination of their wireless service.

Essentially, customers should not conclude that every bill for a telephone service has both deniable and non-deniable charges on it. Prior to withholding payment for any charge on a bill, a customer should verify the status of the charge with the billing company.

After reviewing petitions for reconsideration relating to truth-in-billing and billing format, the FCC on March 29, 2000, released an Order on Reconsideration that reaffirmed the requirement that telephone bills highlight new service providers and prominently display a contact number for inquiries. This requirement is intended to act as a deterrent to slamming and cramming by allowing consumers to more easily identify changes in providers on their bills. The rule does not cover services provided on a per-transaction basis like directory assistance. Changes in a customer's local or long-distance company would be covered.[8]

The Order also adopted proposals to require carriers to use standard industry-wide language and clear descriptions for line item charges identified as resulting from federal regulatory activity. The FCC felt that current presentations of these charges on telephone bills are misleading, inaccurate, and confusing. As a result, through a proceeding announced in the *Federal Register,* the FCC will seek comment from consumer and industry groups concerning standard labels for these charges.[9]

Finally, carriers must prominently display on each bill a toll-free number (or numbers) that customers may use to inquire about or dispute any charge on their bill.

Provisions of this Order not subject to further rulemakings become effective 30 days after the publication of notice of the effective date in the *Federal Register.* That notice was published in the October 12, 1999 *Federal Register* on pages 55163-64. All other principles and guidelines adopted in the Order became effective on November 12, 1999.[10]

The FCC provides consumer information on truth-in-billing at their Web site at [http://www.fcc.gov/Bureaus/Common_Carrier/Factsheets/truebill.html].

The FCC also provides consumer information relating to additional telephone issues at [http://www.fcc.gov/cib/information_directory. html#telephone], [http://www.fcc.gov/cib/consumerfacts/understanding.html] and [http://www.fcc.gov/ccb/consumer_news].

Complaints concerning telephone service may be filed via the FCC Web site at [http://www.fcc.gov/cib/ccformpage.html].

State regulatory authorities may also address telephone billing formats and customer service practices of telephone companies operating in their state.

[8] *Federal Register,* July 13, 2000, p. 43251-58.
[9] *Federal Register*, June 25, 1999, p. 34499-501.
[10] A summary of this Order was published in the *Federal Register* on June 25, 1999, on pages 34488-98.

CHARGES ON LOCAL TELEPHONE BILLS

Local Telephone Service

This is the basic amount that a customer pays for local dialing service, not including any taxes or additional services. State public utility commissions regulate this charge, not the FCC.

The geographic size of a local dialing area and the structure of local dialing service packages vary from company to company and from state to state. Typically, customers may have local telephone service that allows an unlimited number of calls *within* their local dialing area for a flat monthly fee or a service package that allows up to a specific number of local calls during any one month. If a customer exceeds that number of calls, the extra calls are subject to additional charges. Usually, the various local telephone service plans are summarized in the front section of the white pages of a telephone book. In many cases, companies providing local telephone service list the individual component charges that are included in the fee a customer pays for local service. Questions concerning any of these components or the fees charged for each component should be addressed to the company providing local phone service or the state public utility commission. The FCC does not establish or regulate local plan prices or components.

Directory Assistance Charges

Local phone companies, in most cases, assess charges for directory assistance (411) calls. Rates can be as high as $1.25 per call. Charges for 411 calls are not regulated by the FCC. State authorities may regulate these charges. In some states, there are extensive regulations. In other states, the phone companies are given more freedom in assessing 411 fees. Local phone companies may allow a certain number of calls per month to 411 without charging any fee. Above that number, fees are assessed. Customers seeking to avoid these charges should contact their local phone company and ask about the number of free calls that may be permitted; use a phone book; try the Internet; or call a friend to get a number.

Inside Wiring

In some cases, a charge labeled "Inside Wiring" may appear on a customer's bill. This is an *optional* charge that customers may pay to a company for service calls on the wiring *inside* their home. Monthly fees for inside wiring insurance vary from company to company. Inside wiring is owned by the home or building owner.

Customers paying this fee are not charged any additional monies if the company is requested to repair inside wiring. Customers choosing not to pay this fee will be charged by the phone company for any necessary inside wiring repairs requested. Fees charged for inside wiring work vary from company to company. If a customer has an inside wiring problem, there is no requirement to call the phone company. Since the wiring is owned by the home or building owner, any company may be called or the owner may choose to work on the wiring.

Toll Calls

Each telephone customer is permitted (within the parameters of their local dialing plan) to call certain telephone exchanges in their geographic area without incurring any additional charge on their telephone bill. Because of the introduction of new area code overlays, local telephone calls may require seven-digit or 10-digit dialing. Calls made *outside* of a customer's local dialing area, but not going far enough to be classified as long-distance, will incur additional charges on a telephone bill. Local dialing areas are not determined by the FCC. State authorities regulate the local dialing areas in their state and make the determination as to whether calls to certain exchanges are within a specific local dialing area or are toll calls.

Toll calls are often handled by the same company that provides local telephone service to a customer. However, m many states, state authorities have permitted long-distance carriers to compete in the toll call market. Toll call rates can vary substantially depending upon the carrier chosen (in states where such competition is permitted) and other factors. If there are any questions concerning toll call charges or whether a specific exchange is included in a local dialing area, they should be addressed to the company providing local telephone service or state authorities.

Jamming

With the advent of competition in the toll call market, complaints have arisen that some customer accounts are being frozen so that customers cannot use a company competing with their local phone company to complete a toll call. This tactic is referred to as jamming. In cases where this is occurring, customers may be paying more for their toll calls. Customers who feel that they have been jammed and have inquiries about competition in the toll call market should address their inquiries to state authorities.

Sliding

Some customers have also complained that their chosen provider of toll call service has been switched without their permission. This practice has been termed sliding. As with jamming, toll call rates can vary substantially front company to company. Consumers who believe that they have been victims of sliding should contact their chosen toll call provider or state authorities.

Miscellaneous Caller Services

Local telephone companies offer a wide variety of caller services such as: caller ID, call waiting, call forwarding, call rejection, call trace, call return, priority ringing, and voice mail, among many others. Both the types of caller services offered and the charges for these optional services vary from company to company. Charges may include monthly fees or per-use charges. The FCC does not regulate these charges.

Long-Distance Services

Long-distance charges are wholly dependent upon the long-distance company that a consumer chooses as his/her long-distance carrier, the particular calling plan (if any) chosen, and the number and length of calls made during a billing period. Usually, customers designate a specific long-distance company as their primary long-distance carrier. When a customer dials a long distance call by dialing 1+(area code)±telephone number, a telephone switch automatically routes the call to the customer's designated long-distance earner. However, customers are not required to use their designated long-distance carrier to handle any of their long-distance calls. If customers use "dial around" long-distance carriers (reached by dialing the appropriate 10-10-XXX code for a particular company) instead of their chosen long-distance carrier, charges for those calls can also be included on a local bill. In addition, customers may use different types of prepaid phone cards or long-distance company calling cards to complete a long-distance call. Rates and conditions for these cards vary widely, and the charges billed to a card can vary depending upon whether a payphone, business, or residential phone is used.

Many long-distance companies are now billing their residential customers directly instead of billing through a local telephone company. As a result, customers may receive a minimum of two bills for telephone service.

On March 1, 2000, the FCC and the Federal Trade Commission (FTC) issued a joint *Policy Statement* concerning advertising practices relating to long-distance services, especially dial-around (10-10) numbers.[11] The agencies took this action following thousands of complaints from consumers and issued the *Statement* to "... encourage industry to adhere to the standards offered in the joint Policy Statement." According to the FCC and FTC, the *Policy Statement* does not preempt any existing state law.

Suggested guidelines for advertising of long-distance services are:

- All claims must be truthful, non-misleading, and substantiated;
- Carriers should disclose all costs consumers may incur, such as per-call minimum charges, monthly fees, and universal service charges;
- Advertising should disclose any time and/or geographic restriction on the availability of advertised rates;
- The basis for comparative price claims should be disclosed, and only current information used in making claims; and
- Information should be disclosed in a clear and conspicuous manner, and without distracting elements, so that consumers can understand it and make fully informed choices.

---

[11] Joint FCC/FTC Policy Statement For the Advertising of Dial-A round And Other Long-Distance Services To Consumers, Policy Statement, File No. 00-72, FCC 00-72, released March 1, 2000.

The FCC provides Consumer Fact Sheets with tips for lowering your long-distance telephone bill [http://www.fcc.gov/cib/consumerfacts/lowerbill.html] and long-distance shopping [http://www.fcc. gov/cib/consumerfacts/Smartshoppin.html] at its Web site.

Single Bill Fees

Customers who receive a single bill for local and long-distance services may be charged a fee by the long-distance company for this service. The fee is not mandated by the FCC and is not a federal charge. In some cases, customers are informed in advance about the fee. In other cases, no advance notice is given. Should this charge appear on a phone bill, a customer must contact their long-distance company and inquire about separate billing for long-distance calls. The fee will then not apply. Single bill fees are approximately $1.50 per month.

Slamming

Generally, slamming is the unauthorized change of a customer's long-distance service provider.[12] There are existing FCC rules and policies designed to protect telephone customers from this practice, and sections of the Telecommunications Act of 1996 prohibit carriers from changing a customer's long-distance company without following specific verification procedures. On April 13, 2000, the FCC adopted additional rules to combat slamming.[13] As a result, state regulatory agencies will be responsible for resolving slamming disputes. In cases where a state elects not to administer the slamming rules, the FCC will resolve disputes. The new rules also require slammers to compensate both the consumer and the authorized carrier.

Consumers may verify the long-distance carrier connected to their home phone by calling *1-700-555-4141* **from their home phone.** A recording will state the name of the long-distance carrier connected to that line. This is an automated service. Consumers cannot call the 700 number from another location to verify service on their home phone. Calls must be made from the line for which one wishes to verify service.

If there is a problem, customers should contact their local telephone company and chosen carrier and arrange to be switched back to the chosen carrier at no charge. If there was a charge for switching and/or higher rates when slammed, customers have the right to demand a refund. Consumers may also choose to contact their state's Attorney General, public utilities commissions, and/or a consumer protection group or agency.

The FCC provides information on telephone slamming at its Web site at [http://www.fcc.gov/cib/consumerfacts/slamming.html].

Complaints concerning slamming may be filed directly with the FCC:

Federal Communications Commission
Consumer Information Bureau
445 12th Street, S.W.
Washington, D.C. 20554

---

[12] For an overview of the slamming issue, see CRS Issue Brief
[13] *Federal Register,* August 3, 2000, p. 47678-47693.

Complaints may also be filed electronically at the FCC's Consumer Information Bureau Web site [http://www.fcc.gov/cib/ccformpage.html].

Minimum Use Fees

Certain long-distance carriers charge minimum use fees to some of their long-distance customers. In most cases, basic rate customers (those customers who are not on any calling plan) are assessed the charge. However, in some circumstances, customers on calling plans may also be charged. The companies stated that it was necessary to assess the charge because of the expenses of billing, account maintenance, and customer service. Consumer advocates condemned the charge as punishing low volume callers. The fee, if assessed, can be $3 or more per month.

Long-distance carriers may exempt qualifying low-income customers from paying the fee and, usually, long-distance calls made during the month are applied against the fee. If a customer makes $2.50 in long-distance calls during the month, 50 cents will be added to the bill to bring charges up to the $3 minimum. If calls exceed $3, there is no additional fee. Questions about the structure of these fees or company policies concerning the fees should be directed to a customer's long-distance earner.

Consumers should note that a minimum use fee is different from the monthly charge that may be assessed by a particular company's calling plan. A customer might pay $5.95 per month to be on a plan that offers long-distance rates of 10 cents per minute, 24 hours per day. This charge does not increase or decrease regardless of the volume or cost of calls made during the billing period.

Customers who wish to avoid minimum use charges may contact their long-distance carrier and inquire about discount calling plans, switch to a long-distance company that does not charge minimum use fees, or cancel their designated long-distance carrier. Most long-distance calling plans, regardless of the company, may carry basic monthly charges. These charges often exceed $3. Should a customer cancel the designated long-distance carrier, they will still receive *incoming* long-distance calls, but would only be able to make *outgoing* long-distance calls by using dial-around carriers, prepaid calling cards, or cell phones. Customers choosing this option should pay strict attention to the details of the price structure of dial-around or prepaid services. Prices for these methods vary significantly.

CALLS Revisions

Neither the FCC nor the states currently regulate minimum use fees charged by long-distance companies. On July 9, 1999, the FCC announced that it would begin an inquiry into how these fees affect low volume callers.[14] This inquiry was concluded as part of the CALLS proposal adopted on May 31, 2000 (see section Coalition for Affordable Local and Long-Distance Services). While minimum use fees were not abolished, companies agreed to eliminate them or make these fees avoidable through special calling plans with no minimum monthly charge.

---

[14] *Federal Register,* August 5, 1999, p. 42635-37

The FCC provides a series of telecommunications tips at their Web site for choosing a long-distance provider. See [http://www.fcc.gov/marketsense], as well as tips for lowering people's long-distance telephone bills at [http://www.fcc.gov/cib/consumerfacts/lowerbill.html], and long-distance shopping at [http://www.fcc.gov/cib/consumerfacts/Smartshoppin.html].

Internet Access and Long-Distance Charges (Reciprocal Compensation)

Members of Congress and the FCC have been inundated with inquiries concerning the classification of telephone calls to Internet Service Providers (ISP) as long-distance instead of local. Those complaining believed that Congress and/or the FCC were about to enact provisions that would make all calls to ISPs subject to long-distance charges. There were and are no bills in Congress to do this.

The FCC conducted a proceeding at the request of telephone carriers to clarify how local telephone companies should compensate each other (reciprocal compensation) for carrying telephone traffic to ISPs. Essentially, when Telephone Company X (a local phone provider) delivers a local call to the ISP, who has chosen Telephone Company Z to handle its local calls, X pays Z to deliver the call to the TSP. If the ISP calls someone, Z pays X to deliver the call. Charges paid from X to Z or Z to X are based upon the length of time that the call is connected or some other basis determined by X and Z. X and Z enter into an agreement for a specified period of time to compensate each other for carrying calls. This compensation is paid between X and Z and does not involve any charges to the ISP or its customers and has no direct bearing on the fees that an ISP charges its customers.

However, calls to ISPs tend to last a long time, since using the Internet is usually not a speedy endeavor, but outbound calls from ISPs do not (in most cases). Thus, local phone companies like X end up paying a lot more to Z than Z pays to X since the compensation is often based upon the length of time that the call is connected. X and other local phone companies in the same position petitioned the FCC to reconsider the status of these calls and designate them as interstate instead of local Reciprocal compensation applies only to local telephone calls.

On February 25, 1999, the FCC ruled that "... Internet traffic is jurisdictionally mixed and appears to be largely interstate in nature" and in a Notice of Proposed Rulemaking is seeking to determine a federal inter-carrier compensation mechanism.[15] Designation of these calls as interstate by the FCC is a purely jurisdictional designation. Although this ruling means that the structure and method of reciprocal compensation (payments between X and Z) will change. It does not change the status of local calls to ISPs to long-distance for the purposes of billing individual customers. It also does not require an end to reciprocal compensation. The U.S. Court of Appeals for the District of Columbia ruled, on March 24, 2000 that the FCC had not adequately justified its analysis of phone calls to Internet service providers as interstate. The case was sent back to the FCC for further explanation.

---

[15] *Federal Register*, March 24, 10999, p. 14203-6 and 14239-43.

In an Order released on April 19, 2001,[16] the FCC concluded that telecommunications traffic delivered to ISPs is interstate access traffic and is not subject to reciprocal compensation. However, the FCC did not abolish reciprocal compensation. Instead, it established a gradual reduction of the rates over the 2 years following the effective date of the order.

The FCC does not regulate the fees that ISPs charge their customers for Internet access. ISPs construct their own packages of monthly, weekly, hourly, or per-minute charges for their customers.

Additional information on reciprocal compensation is available via the FCC Web site at [http://www.fcc.gov/Bureaus/Common_Carrier/ Factsheets/nominute.html].

Federal Telephone Excise Tax

The federal telephone excise tax first appeared in 1898 as a temporary tax to finance the Spanish-American War. The tax reappeared in 1914 as a tax on long-distance service necessitated by World War I. It has been repealed and reinstated several times since then. The tax was made permanent by the Revenue Reconciliation Act of 1990 (Public Law 101-508) and is currently assessed at a rate of 3% on local and long-distance telephone services. Monies collected from this tax are not kept by the telephone companies but are forwarded to the U.S. Department of the Treasury for general revenue purposes.[17] The Telecommunications Act of 1996 (Public Law 104-104) did not alter this tax. Legislation to repeal this tax was introduced, but did not become law, in the 106th Congress. Similar legislation has been introduced in the 107th Congress.

Recent telephone excise tax collections have been as follows: FYI999 ($5.2 billion), FY1998 ($4.8 billion), EY1997 ($4.7 billion), and FY1996 ($4.2 billion).

Excise Tax on Frequent Flier Miles

Some long-distance companies offer their customers the chance to earn frequent flier miles based on their long-distance calling plan. The long-distance companies purchase the miles from airlines and award them to their customers according to the rules of the promotion being offered. Congress, m the Taxpayer Relief Act of 1997, established an excise tax of 7.5% on the purchase of frequent flier miles. The airlines collect the tax front the companies that purchase the miles. Sonic long-distance companies have chosen to pass all or part of this tax on to their customers who receive the miles. They are free to do so, but are not required to do so. Although the long-distance companies collect this charge via the telephone bill, it is not a telephone-related charge and although the charge is the result of a federal excise tax, it is not related to the 3% federal telephone excise tax mentioned above.

---

[16] U.S. Federal Communications Commission. *In the Matter of... Intercarrier Compensation for ISP-Bound Traffic,* Order on Remand and Report and Order, CC Dockets 96-98 and 99-68, FCC 01-131. Adopted April 18, 2001. Released April 27, 2001. Available via the FCC Web site at [http://www.fcc.gov/ccb7l.

[17] For additional information on the telephone excise tax, see CRS Report RS20119, Telephone Excise Tax, and CRS Report RL30553, *The Federal Excise Tax on Telephone Service: A History,* both by Louis Alan Talley.

Local Number Portability (LNP)

The Telecommunications Act requires implementation of local number portability. LNP permits telephone customers to retain their telephone number even if they switch telephone companies. LNP is being implemented in stages and will initially be available in the 100 largest metropolitan areas. Phone companies reportedly have spent approximately $3 billion to implement LNP.

> **As of February 1, 1999, local phone companies may, but are not required to, assess a monthly charge on customers' telephone bills to recover some of their costs incurred in implementing LNP.**

A monthly charge for LNP may appear on customers bills only in areas where LNP has been implemented. The charge will vary from company to company and region-to-region depending upon the costs incurred to implement LNP. According to various reports, LNP charges that have been assessed have been in the 20 to 60 cents range. In most cases, residential and business customers will be charged the same amount. The charge is permitted to continue for 5 years from the date it first appears, but should not increase during that time.

Since LNP is being implemented in stages, customers in the largest metropolitan areas will probably see the charges first, while customers in other areas will not see any charge until LNP is implemented in their area. Any carrier assessing an LNP end user charge must file a tariff with the FCC.

Wireless and competitive local exchange carriers (CLEC) and long-distance companies have also incurred costs associated with LNP. These companies are not subject to the same restrictions regarding cost recovery and are free to charge their customers as much or as little as they want over a period of time of their choice to recover the costs associated with LNP implementation. As a result, customers may see wireless, CLEC, and long-distance companies assessing an LNP charge also.

On July 1, 1999, following a 5-month investigation, the FCC announced that it had directed several local phone providers to reduce their charges for LNP. According to the FCC, this action will result in a savings of $584 million to consumers.

An FCC Fact Sheet on telephone number portability is available via the FCC Web site at [http://www.fcc.gov/cib/consumerfacts/localport.html].

Universal Service

Section 1 of the Communications Act of 1934, as amended, states that one of the reasons for creation of the FCC is to

"... make available, so far as possible, to all the people of the United States, without discrimination on the basis of race, color, religion, national origin, or sex, a rapid, efficient, Nation-wide, and world-wide wire and radio communication service with adequate facilities at reasonable charges ..."

The Telecommunications Act of 1996 added section 254 (Universal Service) to the Communications Act. This section states that policies for the preservation and advancement of universal service shall be based upon, among other things, quality service at just, reasonable, and affordable rates and that access to advanced telecommunications and information services should be available in all regions of the nation.

The concept of universal service can trace its roots to the turn of the century and the early years of the telephone system in the United States. During these years, a complex system of cross subsidies developed to fund telephone services for all citizens of the United States. Wiring rural areas was much more expensive than wiring urbanized population centers. Profits generated in the urbanized areas were used to subsidize rural wiring. Higher rates were charged for business customers and long-distance calls, enabling lower residential charges for local calling. Later, assistance was provided for low-income households. As the years passed, revenues continued to increase, and these complex cross subsidies enabled the funding of universal service at affordable rates for all citizens.

Up until the present time, telephone bills for the most part did not include itemized charges for universal service. While, technically, all telephone customers have contributed to universal service for decades, such charges were built into the rate system. The companies that currently pay into the universal service mechanism do so based upon their revenues, not according to a specific fee. The FCC has not established rules mandating or forbidding phone companies from itemizing their universal service costs on telephone bills, and there is no specific federal universal service charge that must be charged directly to customers. Presently, phone companies are taking different approaches to itemizing universal service costs on customers' bills. Some phone companies feel that they must pass on the costs of universal service directly to their customers and are itemizing charges on bills to reflect this. Any charge on a phone bill labeled as a "federal universal service charge" or "universal service connectivity charge" or something similar has been added as a specific item by the company issuing the bill. Questions about any such charges should first be directed to that company.

With the divestiture of AT&T in 1984, the expansion of competition, and advances in technology, the structure of the telecommunications industry in the United States began a complex transformation that continues today. No longer was the system of cross subsidies applicable mainly to a single major provider of telephone service. At divestiture, seven "Baby Bells" were created. Now, there are three. The number of local and long-distance providers mushroomed, and the country entered the information age. Telephone service was no longer limited to a wired connection in a home or business. New questions arose relating to the concept of universal service. What type of connections should be included? Who should contribute to a universal service mechanism? How much should they pay? How should they pay?

As a result of the Telecommunications Act of 1996, the FCC attempted to answer some of these questions. In its May 7, 1997, universal service and access reform decisions, the agency, in compliance with the provisions of the 1996 Act, expanded the field of entities eligible for universal service to include schools and libraries (known as the "E-rate"[18]) and

---

[18] For more detailed information on the E-rate, see CRS Issue Brief IB98040, Telecommunications Discounts for Schools and Libraries, and CRS Report 98-604, *E-Rate for School: Telecommunications Discounts Through the Universal Service Fund.*

rural health care providers. The pool of companies paying to fund universal service was enlarged and access charges were restructured.

On May 8, 2001, the FCC initiated a review of the way that telecommunications carriers contribute to the universal service fund. Part of the review will be to ensure that the carriers' contributions are recovered (from customers) fairly, accurately, and equitably.

For extensive information on the FCC's actions relating to universal service and copies of documents see [http://www. fcc.gov/ccb/universal_service/welcome.html], as well as the FCC Consumer Fact Sheets on universal service at [http://www.fcc.gov/cib/information_directory. html#telephone].

The universal service fund is administered by the National Exchange Carrier Association (NECA), who can be reached via their Web site [http://www.neca.org] or at:

National Exchange Carrier Association
80 South Jefferson Road
Whippany, NJ 07981-1009
(800) 228-8597

NECAs subsidiary, the Universal Service Administration Company (USAC), provides detailed information on the various components of the universal service mechanism at [http://www.universalservice.org].

Local Taxes

The county, city, or state in which an individual lives often has its own tax on telephone service. Local taxes may be much higher than the federal excise tax and can exceed 20%. Local taxes may include: franchise, gross receipts, state sales, local sales, municipal, special district, or earnings taxes, and state mandated universal service surcharges.[19] Some or all of these charges may appear on a customer's telephone bill. Questions about these taxes should be directed to local phone companies, state public utility commissions, or local tax authorities.

Interstate Tax Surcharge

This charge, also known as a gross receipts tax, applies to interstate revenues generated by long-distance telephone companies within an individual state. It is not a federal tax. State or local tax authorities can provide information on this tax.

---

[19] *Telecommunications Tax Policies: Implications for the Digital Age,* National Governors' Association, February 2000.

911 Charges

Local government authorities are responsible for the construction and maintenance of 911 emergency calling systems within a state. Any 911 charges or taxes appearing on a bill are dependent upon a local government's actions relative to 911 and will vary from locale to locale. Implementation of an enhanced 911 (E911) system is underway for wireless service providers. As a result, customers of cellular and personal communications services and other wireless service companies may see 911 charges on their bills.

Relay Center Surcharges

Also known as Telecommunications Relay Services (TRS), this charge is used to provide operator-assisted telecommunications services for people with hearing or speech disabilities. Costs for intrastate TRS services are paid by the states. Costs for interstate TRS services arc borne by the Interstate TRS fund, administered by the National Exchange Carrier Association and funded by all interstate carriers. The NECA collects funds from approximately 3,000 companies based on their interstate revenues. Charges on customers' bills are usually a few cents per telephone line. TRS services are required by Title IV of the Americans with Disabilities Act (Public Law 101-336). For additional information on TRS, see [http://www.fcc.gov/cib/dro/trs.html].

Cramming

Customers who cannot determine what a specific charge is for might have been "crammed." Cramming refers to the inclusion of unauthorized or possibly illegal charges that appear on a customer's bill. An amount might be labeled as "monthly fee," "membership," or "information service." Contact should be made with the local telephone company or bill provider to obtain the name, address, and phone number of the company for whom they are collecting the fee in question. Consumers sho7uld request that the charge be removed from the bill fithey believe they are a victim of cramming. Since the local phone company is usually only acting as a billing agent for a company, they cannot resolve individual disputes. However, they should be made aware of the situation.

Complaints concerning questionable charges for calls placed within a customer's state should be directed to a local consumer office or the state public utility commission and the company that initiated the charge in question.

If the charges involve information services (900 numbers, psychic hotlines, etc.), not telephone services, a customer may register a complaint with and obtain information from:

Federal Trade Commission
Consumer Response Center, Room 130
600 Pennsylvania Avenue, N.W.
Washington, D.C. 20580
Toll free: (877) 382-4357
FTC Web site: [http://www.ftc.gov]

Should the Complaint involve telephone-related issues, interstate or international services, or charges, a complaint may be registered in writing with the FCC:

Federal Communications Commission
Consumer Information Bureau
445 12th Street, S.W.
Washington, D.C. 20554
Toll free: (888)-CALL-FCC (225-5322)

The FCCis currently conducting an inquiry into invalid and unclear charges on telephone bills. The FCC provides further information at their Web site [http://www.fcc.gov/Bureaus/Common_Carrier/Fact sheets/cramming.html].

Information on the FCC's anti-cramming best practices guidelines is also available at the FCC Web site [http://www.fce.gov/Bureaus/Common-Carrier/Other/cramming/ermmning.html].

A Fact Sheet on cramming is available at [http://www. fcc. gov/cib/consumerfacts/cramming.html].

Internet Cramming

Recently, a cramming scam that targets small businesses, religious groups, charities, foundations, or any small organization desiring an Internet presence has generated thousands of complaints. Companies, usually through some type of telemarketing operation, will contact consumers and offer a "free trial" for the design and maintenance of a Web site. In many cases, such companies fail to disclose that, unless the free trial is specifically canceled by the consumer, a monthly fee (for continued maintenance of the Web site) will be collected and charged to a customer's telephone bill. In some cases, even when the free trial is canceled by the customer, the charges continue to appear on the customer's phone bills.

The FTC has filed Internet cramming cases against various companies and provides information at their Web site. Complaints may be filed at the FTC Web site or by contacting the FTC at the address or phone number listed above.

CHARGES ON WIRELESS TELEPHONE BILLS

Many of the charges that appear on local (wired) telephone bills also appear on bills issued by wireless companies. Most notably: monthly service charges, itemized call charges (local and long-distance, depending upon the calling plan a customer chooses), federal telephone excise tax, 911 fees, any applicable state or local taxes, universal service charges, LNP, and TRS fees. As with local phone bills, wording used on wireless bills can vary from company to company. Also, wireless companies are not subject to the same truth-in-billing and billing format principles discussed earlier in this report. However, the FCC is currently conducting an inquiry into whether its truth-in-billing requirements should apply to wireless carriers and what uniform labels should be used to identify charges resulting from federal action.[20]

Neither SLC nor PICC charges appear on wireless bills. However, some wireless companies use the wording "Interconnect/Landline Charges," or "Landline Connection Fee" on their bills. This charge is not a result of the FCC access charge mechanism. It is a charge assessed by a local phone company to connect a wireless call through their network to the called party. The charge varies company-to-company and calling plan to calling plan.

Although many charges on wireless bills are similar to or the same as those on wireline bills, there is a series of charges that are particular to wireless. For instance, wireless customers in many cases are billed for both incoming and outgoing calls, and wireless companies may start billing for a call as soon as the send button is pressed. Unlike wireline calls, a charge may appear on a wireless bill for a call that did not go through. In addition, wireless companies often bill by the minute, not by the second. As a result, a call lasting 61 seconds may be billed as a two-minute call. Also, there may be substantial termination fees assessed on customers who try to cancel their service before a certain date, and there are usually geographic limits for coverage provided by different companies and plans.

The FCC has no regulatory authority over these various fees or billing practices. Essentially, when customer signs up for wireless telephone services, they have entered into a contractual agreement with a company. Customers must read the contract carefully and fully understand the terms. In cases where customers believe they have been misled or have been the victims of some questionable business practice, they may wish to contact the office of the attorney general in their state.

FEDERAL COMMUNICATIONS COMMISSION

The FCC Common Carrier Bureau has developed a Web site devoted to various telephone-related issues. The site includes several FCC fact sheets on specific telephone-related issues and summaries of enforcement actions, and it allows consumers to file complaints via the Web.

See [http://www.fcc.gov/cib/information_directory.html#telephone]

Federal Communications Commission
Consumer Information Bureau
445 12th Street, S.W.
Washington, D.C. 20554
Toll free: (888)-CALL-FCC (225-5322)

The FCC Web site provides a list of main, complaint, and in-state, toll-free telephone numbers for the telecommunications regulatory authorities in each state. The list is available at [http://www.fcc.gov/ccb/consumer_news/state_puc.html]. No mailing addresses are provided.

---

[20] *Federal Register*, June 25, 1999, p. 34499-501.

NATIONAL ASSOCIATION OF
REGULATORY UTILITY COMMISSIONERS

The National Association of Regulatory Utility Commissioners (NARUC) is an organization of state and federal regulatory commissioners having jurisdiction over public utilities. Individual state public utility commissions (PUC) may provide assistance to consumers concerning telephone bills and any state laws and regulations that may apply to companies providing telephone services within the state.

Web site connections and addresses for state PUCs are available through the NARUC Web site at [http://www.naruc.org]. Also, NARUC can be contacted at the following address:

National Association of Regulatory Utility Commissioners
1101 Vermont Avenue, N.W., Suite 200
Washington, D.C. 20005
(202) 898-2200

Chapter 4

WIRELESS TECHNOLOGY AND SPECTRUM DEMAND: THIRD GENERATION (3G) AND BEYOND

Linda K. Moore

SUMMARY

Advances in wireless telecommunications technology are converging with Internet technology to foster new generations of applications and services. Presently, the United States and other countries are moving to a third generation (3G) of mobile telephony. The defining feature of 3G is that data transmission speeds are significantly faster than for prevailing technology. Many of he new wireless services being introduced or on the drawing boards incorporate Internet-like features that make heavy demands on bandwidth capacity. Industry experts have observed that more efficient uses of spectrum must be developed to meet growing demand.

The U .S. Congress and Federal Departments and Agencies are among the organizations examining the impact that 3G and future technologies will have on bandwidth demand and spectrum allocation. Several initiatives are underway including proposed legislation to postpone, from September 30, 2002 to September 30, 2004, a planned auction of spectrum intended for 3G use.

WIRELESS TECHNOLOGY: DEVELOPMENT AND DEMAND

Spectrum bandwidth is a finite resource. Commercial wireless communications currently rely on bandwidth within a narrow range, a "sweet spot."[1] American competitiveness in this key technology is constrained by the amount of useful bandwidth that is available. This constraint is both specific, in the inherent finiteness of spectrum, and relative, in comparison to spectrum available for commercial use in other countries.

---

[1] Martin Cooper, Chairman and CEO, ArrayComm, Inc., San Jose, CA, describes the "sweet spot" as 2000 megahertz (MHZ) of bandwidth in approximately the 7000-2700 MHZ range.

Wireless communications services have grown significantly worldwide, and explosively in some countries. Demand for wireless telephony in the United States has almost tripled since 1995; the number of subscribers had surpassed 123 million by mid-October 2001.[2] In approximately the same time frame, use of the Internet expanded dramatically from an arcane tool for specialized research to a popularized, user-friendly service providing near instant access to information and entertainment. Internet services delivered by telephone or cable connections were quickly followed by wireless delivery to portable computers, personal digital assistants (PDAs), telematics systems in cars, or mobile telephones. Business and consumer demand for new, advanced wireless services is considered by many to be a potential engine for future growth in the American global economies.

Technology Development

Mobile communications became generally available to businesses and consumers in the 1980s. This "first generation" technology – still in use – is analog, the prevailing telecommunications technology of the time. Second generation (2G) wireless devices are characterized by digitized systems that provide better delivery of voice and small amounts of data, such as caller ID. Also, digital wireless is significantly more efficient in its use of spectrum than is analog cellular technology.[3]

The next major advance in mobile technology is referred to as the "third generation" – "3G" – because it represents significant advances over the analog and digital services that characterize current cellular phone technology. The dramatic increase in communications speed is the most important technical feature of 3G.[4] Higher transmission speeds are essential for robust Internet connections. Projected speeds of 2 megabits per second have been realized in laboratory conditions but many industry observers have noted that real-world transmission speeds where 3G is in service are well below 384 kilobits per second.

Many hypothesize that 3G will be an interim technology and that new developments in wireless and Internet technologies will advance to another level within several years. An example of this is the revival of time division duplex (TDD) transmission technology. Some analysts in Europe predict that TDD will replace current standards for delivering 3G, improving both bandwidth capacity and the quality of Internet services delivery.[5]

---

[2] Cellular Telecommunications and Internet Association (CTIA), web site [http://www.wow-com.com].
[3] As much as 18 times more efficient, depending on the technologies being compared, according to protocols.com [http://www.protocols.com].
[4] The Federal Communications Commissions (FCC) identifies key service attributes and capabilities of 3G as the following: capability to support circuit and packet data at high bit rates; interoperability and roaming; common billing and user profiles; capability to determine and report geographic position of mobiles; support of multimedia services; and capabilities such as "bandwidth on demand." 3G speeds are: 144 kilobits per second at vehicular traffic speeds; 384 kilobits for pedestrian traffic; 2 megabits or higher for indoor traffic, [http://www.fee.gov/3G].
[5] *The Financial Times*, "Comeback for an Older Technology," November 15, 2000.

The Role of Technology in Spectrum Management

In order to deploy 3G and other new technologies, telecommunications carriers and their suppliers are seeking effective strategies to move to new standards, upgrade infrastructure, and develop software for new services. This migration path includes decisions about using spectrum.

Radio frequency (RF) spectrum is used for all wireless communications. It is managed by the FCC for commercial use and, for federal government use, by the National Telecommunications and Information Administration (NTIA). International use is facilitated by numerous bilateral and multilateral agreements covering most aspects of usage, including mobile telephony. Spectrum is typically measured in cycles per second, or hertz.[6] Currently, spectrum below 3000 megahertz (MHZ) is used for commercial military and navigational purposes. Commercial uses for mobile telephony above 3000 MHZ (3 GHZ) are not considered feasible with existing terrestrial wireless technology.[7]

Spectrum is segmented into bands. Developments in technology have in the past facilitated the more efficient use of bandwidth within a given portion of the spectrum. New technologies for terrestrial wireless, such as Software-Defined Radio (SDR) and "smart" antennae, are being explored and implemented to increase the efficiency of spectrum frequencies.

Efficiencies in wireless technology can be achieved in every aspect of the system, what is called a "value chain." For example, SDR chips improve efficiency in the handset, smart antennae increase the efficiency of transmission towers, compression technologies send more data in less bandwidth, and TDD techniques use bandwidth more effectively.

PUBLIC POLICY AND 3G

International Agreements on 3G

International agreements that coordinate and enable global telecommunications are negotiated under the aegis of the International Telecommunications Union (ITU), a specialized agency of the United Nations. At ITU conferences in 1992 and 2000, resolutions were passed regarding the use of spectrum for the International Mobile Telecommunications-2000 (IMT-2000) initiative. IMT-2000 is "the ITU vision of global mobile access ... intended to provide telecommunications services on a worldwide scale regardless of location, network, or terminal used."[8] The applications for IMT-2000 will come from 3G technology, although 3G is potentially more far-reaching than the current ITU concept.

Delegates to the ITU World Radio Conference in 2000 (WRC-2000) agreed that harmonized worldwide bands for IMT-2000 were desirable in order to achieve global roaming and economies of scale. Resolutions voted by WRC-2000 delegates encouraged nations to make available some part of one or more of the three spectrum bands identified in

---

[6] One million hertz = 1 megahertz (MHZ); 1 billion hertz = 1 gigahertz (GHZ).
[7] The FCC, for example, is limiting its consideration of bandwidth available for 3G to frequencies below 3 GHZ.
[8] Resolution 223, "Final Acts of WRC-2000", ITU, Geneva, Switzerland, [http://www.itu.int/publications].

committee (806-960 MHZ, 1710-1885 MHZ, and 2500-2690 MHZ) for use as harmonized spectrum.

Harmonized Spectrum

The applications of wireless technology are tied to spectrum. Infrastructure, such as towers, relay stations, and handsets, must be able to provide communications along pre-designated frequencies. The value of harmonization is to provide common bands of spectrum dedicated to 3G technology worldwide. This makes it easier for carriers to cover large geographical areas and for the telecommunications industry to develop 3G hardware and software. Many industry observers, however, believe that WRC-2000 did not evaluate the practical considerations of achieving global roaming capabilities and economies of scale through harmonization. They argue that countries like China and Brazil are using spectrum to develop 3G technology in bandwidths not covered by the WRC-2000 resolution, and that global roaming exists today without the benefit of harmonized spectrum. Market demand is perceived by many as sufficient large to provide desirable levels of scale and scope. In planning the American migration from existing wireless to 3G, some experts believe that spectrum in other frequency bands would be the optimal choice.

POLICY ISSUES IN THE UNITED STATES

In a memorandum dated October 13, 2000 to Executive Departments and Agencies, President Clinton followed up on the WRC-2000 delegation's actions by directing the Secretary of Commerce to work with the FCC, in coordination with the NTIA, to prepare studies on allocating bandwidth for harmonized spectrum. President Clinton set July 2001 as the date by which the FCC should have identified spectrum to be used to meet the WRC-2000 objectives, and September 30, 2002 as the deadline to auction licenses for this spectrum. The memo also directed the Secretaries of State, Defense, the Treasury, Transportation and other departments or agencies using spectrum identified by WRC-2000 for third-generation wireless services to cooperate in the effort.

Reports on Spectrum Use

In March 2001, the NTIA and the FCC issued reports, respectively, on 1710-1850 MHZ and 2500-2690 MHZ use.[9] In its report, NTIA divided the band into two segments: the 1710-1755 MHZ band already scheduled to be made available for commercial use[10], and the 1755-1850 MHZ band occupied by the Department of Defense (DOD) and 13 other government

---

[9] "The Potential for Accommodating Third-generation Mobile Systems in the 1710-1850 MHZ Band," Final Report, March 2001, U.S. Department of Commerce, NTIA, [http://www.ntia.doc.gov/reports]; and "Spectrum Study of the 2500-2690 MHZ Band," Final Report, March 30, 2001, FCC, [http://www.fcc.gov/3G].

[10] The "Balanced Budget Act" of 1997 directed the FCC to auction bandwidth from the 1710-1755MHZ not later than September 30, 2002.

agencies. In particular, the report addressed the issue of reallocating spectrum now used by the DOD. The DOD also issued a report on the subject, which reached different conclusions than those of the NTIA.[11]

The report prepared by the FCC covered bandwidth used primarily by Fixed Service operators for Multipoint Distributions Service (MDS), Multichannel Multipoint Distribution Service (MMDS) and Instructional TV Fixed Service (ITFS). On August 9, 2001, at a public hearing, the Chairmans of the FCC pledged to take action by the end of the month regarding the possible reallocation of the 2500-2690 MHZ spectrum frequency. At the same meeting, the FCC adopted a Memorandum Opinion and Order and Further Notice of Proposed Rulemaking[12] seeking comment on the reallocation of additional spectrum for advanced wireless services. These bands are: 1910-1930 MHZ; 1990-2025 MHZ; 2150-2160 MHZ; 2165-2000 MHZ; and 2390-2400 MHZ. Subsequently, the FCC adopted a First Report and Order and Memorandum Opinion and Order[13] adding a mobile allocation to the 2500-2690 MHZ band. The FCC will rely on market forces to determine the "highest and best use" on the band, with incumbent licensees in this band responding with, for example, changes in technology to exploit the future potential of the bandwidth while continuing to support educational TV and other current uses.

In August 2001, the General Accounting Office published a report on Defense Spectrum Management[14] that concluded: "Spectrum decisions based on either the DOD or the industry study[15] of the 1755 to 1850 MHZ band would be premature at this time. Neither study contains adequate information to make reallocation decisions."

Further Study

On June 26, 2001, the Chairman of the FCC wrote the Secretary of Commerce suggesting that more time be allowed for the effort to identify appropriate spectrum and that the deadline for the proposed auction be extended. In response, the Secretary directed the Acting Director of the NTIA to work with the FCC to develop a new plan for the selection of 3G spectrum, with the cooperation of the National Security Council and the DOD, among others.

The NTIA announced the new plan on October 25, 2001, prepared with the FCC, the DOD and other Executive Branch agencies. An interagency assessment is underway that will evaluate options and time lines for making more spectrum available for commercial use in two bands without reducing the effectiveness of current and planned services already in use. The bands identified for assessment are for 1710-1770 and 2110-2170 MHZ. Use of the 1770-1850 MHZ band is not part of the assessment. The study is scheduled for completion by the Federal government in late spring 32002. After the assessment is completed the NTIA will coordinate, first with other Executive Branch Agencies and then with the FCC, to reach

---

[11] "Investigation of the Feasibility of Accommodating the International Mobile Telecommunications (IMT) 2000 Within the 1755-1850 MHZ Band," 9 February 2001, Department of Defense.
[12] FCC 01-224.
[13] FCC 01-256, September 6, 2001, released September 24, 2001.
[14] "Defense Spectrum Management," GAO-02-795.
[15] Industry Association Group on Identification of Spectrum for 3G Services, the Group is a consortium of the CTIA, the Telecommunications Industry Association and the Personal Communications Industry Association.

solutions for allocating spectrum.[16] To accommodate the new schedule, legislation has been proposed that would extend the auction deadline to September 30, 2004.[17]

DOD and FCC Activity

The Department of Defense is proceeding with a project to improve spectral efficiency, the "Next Generation (XG) Communication Program." The program is being conducted jointly by the Defense Advanced Research Projects Agency (DARPA) and the Air Force Research Lab.[18] The Federal Communications Commission has participated in discussions and will work in partnership with DARPA to facilitate technology transfers from the military for commercial use.[19] The Office of Spectrum Analysis & Management, Defense Information Systems Agency, is also involved in the initiative.

Congress

Hearings on laws and policies governing 3G, spectrum allocation, and spectrum frequency management were held by the House Committee on Energy and Commerce, Subcommittee on Telecommunications and the Internet, on July 24, 2001. The topic was addressed, on July 31, 2001, by the Senate Committee on Commerce, Science and Transportation, Subcommittee on Communication. In the current climate of re-evaluation of spectrum policy and bandwidth needs for public services and defense, laws that might be considered for possible amendment include: Communications Act of 1934; Telecommunications Act of 1996; Wireless Communications and Public Safety Act of 1999; Balanced Budget Act of 1997; and FY2000 Defense Authorization Act.

ISSUES FOR CONGRESS

The continued growth in demand for bandwidth for private and public sector use has prompted Congress to review the policies and laws that guide the management of this resource. Among the questions being posed are:

- What law, policies and rulings for spectrum allocation would best meet the sometimes conflicting objectives of protecting consumers, fostering new technology, encouraging efficiency, bolstering international competitiveness, and promoting competition, fairness, and access in domestic markets?
- To what extent does the United States support efforts for international cooperation in developing 3G technology?

---

[16] NTIA Statement at [http://www.ntia.doc.gov/ntiahome/threeg/3gplan]
[17] Letter to The Honorable Richard B. Cheney from Donald L. Evans, September 6, 2001.
[18] Program details are at [http://www.eps.gov/spg/USAF/AFMC/AFRLRRS/]
[19] Press conference remarks by Michael K. Powell, Chairman, FCC, "Digital Broadband Migration," October 23, 2001.

- What are the actions Congress might take to encourage more efficient use of spectrum?
- What effect, if any, do spectrum allocation decisions have on the robustness of America's telecommunications infrastructure?
- What are the trade-off's between commercial needs and those of defense and public safety organizations? What role, if any, might congress have in adjudicating the issue?
- How is the FCC fulfilling its mandates as described in the above-noted laws?

Chapter 5

Wireless Legislation: Wireless Privacy Enhancement Act (H.R. 514) and the Wireless Communications and Public Safety Act (H.R. 438)

Charles Doyle

Summary

The Wireless Privacy Enhancement Act (H.R. 514) and the Wireless Communications and Public Safety Act (H.R. 438) are designed to make emergency telephone service more effective and to ensure the privacy of cellular telephone communications. In doing so, they may also encourage the development and competitiveness of the wireless communications industry.

H.R. 514, as introduced, reemphasizes the criminal proscriptions against the use of scanners to eavesdrop on cellular telephone conversation. H.R. 438 calls for the establishment of a universal emergency telephone number, grants wireless communications service providers with a level of immunity from civil suits comparable to that other communications carrier enjoy under state law; and provides authority for communications carriers to provide authorities with mobile phone customer location information.

Features of H.R. 514

H.R. 514 arose out of concerns that surfaced when a Florida couple used a scanner to eavesdrop on the cellular conversation of a Member of Congress. The couple subsequently pled guilty to a violation of the Electronic Communications Privacy Act, 18 U.S.C. 2510-2522. Later civil litigation was dismissed on First Amendment, free speech grounds.

The bill grants the Federal Communications Commission (FCC) reinforced powers to condition approval of scanners so as to encumber their use for unlawful eavesdropping, proposed 47 U.S.C. 302. It also amends the Communications Act so that it provides parallels rather than supplementary protection for misconduct outlawed under the Electronic Communications Privacy Act, proposed 47 U.S.C. 607. Its disclosure prohibitions at their perimeter may encounter the same first amendment questions that arose out of the activities of the Florida couple, *cf., Smith v. Daily Mail Publishing Co.*, 443 U.S. 97, 103 (1979); *Florida Star v. B.J.F.*, 491 U.S. 524, 533 (1989).

FEATURES OF H.R. 438

H.R. 438, as introduced, permits the FCC to designate 911 as the universal emergency telephone number within the United States for both wireline and wireless telephone service, proposed 47 U.S.C. 251(e). The immunity comparability granted wireless service provider "includes" protection from liability arising out of development, design, installation, operation, maintenance, performance, and provision of service (to the extent not overridden within two years of enactment by state law), H.R. 438, §4. It likewise supplies an equivalent immunity both for liability arising out of the disclosure of subscriber information to authorities and out of difficulties associated with providing emergency service, H.R. 438, §4. The subscriber information disclosure provisions are cast to limit their use to emergency situations, proposed 47 U.S.C. 222.

LEGISLATIVE ACT TO DATE

The Commerce Committee's Subcommittee on Telecommunications, Trade, and Consumer Protection held hearings on the bills in early February at which it heard testimony from the administration, the FCC, local police, the wireless communications industry, and privacy groups. It referred the bills to the full Committee, which ordered them, reported out without amendment on February 11, 1999. The Committee's report is not yet available.

Chapter 6

THE PRIVATIZATION OF COMSAT, INTELSAT, AND INMARSAT: ISSUES FOR CONGRESS

Erin C. Hatch

SUMMARY

International satellite communications have changed significantly in the three decades following the first commercial operations. When communications satellites were developed, economic and technological risks—as well as congressional concerns about potential monopolization by AT&T—necessitated the involvement of both government and private entities. The *Communications Satellite Act of 1962* created the Communications Satellite Corporation (Comsat), a private company that served for many years as the only U.S. provider of satellite communications. An international treaty organization, the International Telecommunications Satellite Organization (Intelsat), was created to set up and maintain a global commercial telecommunications satellite system. The International Maritime Satellite Organization (Inmarsat), another treaty organization, was established to provide communications for commercial, distress, and safety applications for ships at sea.

However, new technologies and applications, and increases in demand have made satellite communications economically feasible for industry to develop commercially. As a result, increased competition in international satellite communications led many to question the organizational structure and market dominance of Intelsat and Inmarsat, and the statutes and regulatory provisions that provide Comsat market advantages. Inmarsat has already become a private company, and Intelsat has begun to privatize by creating a spin-off private company, New Skies. However, until September 1999, U.S. companies could purchase access to Intelsat's satellites only via Comsat. For these and other reasons, some argued that Inmarsat, Intelsat, and Comsat continue to receive privileges and immunities, giving them unfair advantages over other private companies. These parties want to ensure that the organizations privatize in a pro-competitive manner by using access to the U.S. market and Comsat's role as U.S. signatory to Intelsat and Inmarsat as leverage.

Following initial efforts in the 105th Congress, the Senate passed S. 376 (Burns), the *Open-market Reorganization for the Betterment of International Telecommunications*

(ORBIT) Act, in July 1999. The *Communications Satellite Competition and Privatization Act of 1999,* I-I.R. 3261 (Bliley), was introduced in November 1999. The House passed S. 376 by a voice vote with H.R. 3261 as an amendment in the nature of a substitute. However, there were several controversial differences between the bills. Most notably, the House version set stricter standards for deregulating Intelsat, and was more favorable to startup satellite companies, such as PanAmSat and Loral CyberStar. In addition, the Administration and two members of the Commission of the European Union expressed opposition to a preliminary version of the conference bill, citing potential violations of U.S. international agreements, such as those made to the World Trade Organization (WTO). These differences were ultimately addressed within conference, and the conference report (H.Rept. 106-509) passed the Senate on March 2, 2000, and the House on March 9, 2000. On March 17, 2000, the President signed the *ORBIT Act* into law (P.L. 106-180). The FCC and other relevant federal agencies have begun to implement some of the Act's provisions.

INTRODUCTION

Communications satellites receive signals from Earth station antennas, amplify them, and then relay the signals back to another antenna on Earth. As communications satellites are insensitive to terrestrial distances, they provide opportunities for linking remote areas and locations lacking ground communications. In addition, ever increasing satellite capacity and decreasing costs have made communications satellites attractive alternatives to more traditional cable communications lines. They are now utilized for a wide range of voice, data, and video transmissions, and represent by far the most substantial commercial exploitation of space exploration.

However, when communications satellites were first being developed, economic and technological risks—as well as congressional concerns about potential monopolization by American Telegraph & Telephone (AT&T)—necessitated the involvement of both government and private entities, resulting in the *Communications Satellite Act of 1962.*[1] This Act created the Communications Satellite Corporation (Comsat), a private company that served for many years as the only U.S. provider of satellite telecommunications. Subsequent to Comsat's incorporation, the International Telecommunications Satellite Organization (Intelsat) and the International Maritime Satellite Organization (Inmarsat) were established as international treaty organizations. Intelsat was created to set up and maintain a global commercial telecommunications satellite system, and represented the first attempt to exploit commercial space technology on an international basis. Similar to Intelsat, Inmarsat began as a joint cooperative venture between governments and their representative signatories. Inmarsat's original purpose was to provide communications for commercial, distress and safety applications for ships at sea.

Since the creation of these three entities, market, regulatory, and technological changes in the international satellite industry have changed the competitive playing field upon which satellite organizations and corporations provide services to consumers. The resulting increased competition led many to question the organizational structure and market dominance of Intelsat and Inmarsat, and the statutes and regulatory provisions that provide

Comsat market advantages. All three parties responded to changes in the industrial environment by announcing their plans to transition into private companies.[2] Inmarsat became a private company, and Intelsat began the transition with the creation of a spin-off private company, New Skies. In addition, Lockheed Martin announced plans to acquire Comsat, and, following approval in September 1999 by the Federal Communications Commission (FCC), purchased 49% of Comsat's shares.

However, some legislators maintained that the privatization processes were not occurring quickly enough. These individuals asserted that Inmarsat, Intelsat, and Comsat continued to enjoy privileges and immunities, giving them unfair advantages over other private companies. Others stated that Intelsat and Inmarsat had achieved their missions and may no longer be necessary; or, at the very least, that the U.S. had nothing to gain by protecting or promoting either organization.[3] Still others believed that the organizations were still necessary in some form to guarantee safety at sea and services such as international telephone and data transmission, especially for developing countries.[4] Congress considered legislation that sought to amend the *Satellite Act* so as to help ensure that the organizations privatize in a pro-competitive manner, using access to the U.S. market and Comsat's role as U.S. signatory as leverage. In addition, as the *Satellite Act* sets ownership limits on Comsat, the Act had to be changed to in order to permit further acquisition by Lockheed Martin.

Though three related bills were introduced during the second session of the 105th Congress (H.R. 1872, S.1328, and S. 2365*)*, only one (H.R. 1872) passed the House. During the first session of the 106th Congress, two bills were introduced (S. 376 and H.R. 3261), one of which (S. 376) passed both Houses. The House-passed version was based on the text of H.R. 3261, and was significantly different from the Senate-passed version. The most notable differences included "direct access" provisions, differences in privatization deadlines, and criteria for authorizing access to the U.S. market.

These differences were ultimately addressed within conference. The compromise legislation includes provisions to ensure that the privatized successors to Intelsat and Inmarsat be independent of signatories to either organization. If Intelsat and Inmarsat do not privatize according to the legislation's specific criteria and by a certain date, the legislation limits their access to the U.S. market. The legislation gives Intelsat until April 1, 2001, and Inmarsat until April 1, 2000, to accomplish full privatization. It also requires the FCC to limit, deny, or revoke authority for either organization to provide "non-core" services to the U.S. market if the FCC determines that the organization has not privatized by these dates. Finally, the legislation adds a number of deregulatory measures designed to ensure that all U.S. satellite service providers compete as efficiently as possible within the U.S. satellite marketplace.

The conference report (H.Rept. 106-509) passed the Senate by unanimous consent on March 2, 2000, and the House by a voice vote on March 9, 2000. On March 17, 2000, the President signed the *ORBIT Act* into law (P.L. 106-180). While signing the Act, President

---

[1] Public Law 87-624; 47 U.S.C. §701 *et seq.* The Act is hereafter referred to as the *Satellite Act.*

[2] In announcing their plans to privatize, both Intelsat and Inmarsat have set deadlines for the completion of the privatization process, including an initial public stock offering (IPO). As the conduct, precise timing and other details of offerings are decided by the organizations' board of directors and management, however, any future public stock offering is likely to be subject to market conditions at the time.

[3] Milton Mueller, "Intelsat and the Separate System Policy: Toward Competitive International Telecommunications," *Policy Analysis*, No. 150, (Washington: CATO Institute, March 21, 1991).

[4] Leland L. Johnson, "The Future of Intelsat in a competitive Environment," *RAND Publication Series* (Santa Monica, CA: The RAND Corporation, December 1988).

Clinton raised concerns about incompatibility between some of the Act's provisions and U.S. international obligations, such as the World Trade Organization (WTO) membership requirements.

Since passage of the *ORBIT Act,* the FCC and other relevant federal agencies have begun to implement its provisions. At the same time, some of the corporations and organizations most affected by the Act have also begun or continued to incorporate changes in their own structures and policies, some directly related to the Act. Most notably, the FCC granted permission to "directly access" Intelsat to at least three U.S. telecommunications providers, and requested information and comments regarding whether sufficient opportunities exist to access Intelsat's space-based services. The Department of Commerce also sought comments on the progress of Intelsat and Inmarsat signatories in providing full and open competition in satellite markets. Lockheed Martin acquired the remaining 51% stake of Comsat through a one-for-one, tax-free exchange of stock. At the same time, Intelsat's private spin-off, New Skies, asked the FCC for a 6-month extension to the *ORBIT Act's* privatization deadline. Finally, the FCC approved Intelsat's request to become a U.S. licensee for 17 existing and 10 planned satellites. And, some U.S. telecommunications providers have filed applications with the FCC to use Inmarsat's "non core" services following its full privatization. (For more information about these and other events, see the "Most Recent Developments" section later in this report.)

This report summarizes the history and development of today's international satellite communications industry, and the organizational changes within Intelsat, Inmarsat, and Comsat. In addition, the report examines Congress's efforts to affect the continuing development of these three satellite organizations, the issues concerning specific legislative provisions, the current status of this legislation, and the most recent developments since the passage of related legislation. Finally, Appendix A provides a brief history of the creation of all three satellite communications entities.

MARKET, TECHNOLOGICAL, AND ORGANIZATIONAL DEVELOPMENTS

The international satellite communications industry has changed significantly in the three decades following the first commercial satellite operations. The most significant change has been the healthy rate of increased traffic and revenue in the communications satellite market. Key technological innovations have provided for much of this growth, including advances such as the use of more frequency hands and large, high-performance antennas on board satellites.[5] In addition, the range of space communications services has evolved from international and domestic fixed satellite services to mobile satellite services, broadcast satellite services, and even inter-satellite links.

New technologies, combined with new applications and increases in demand, have made satellite communications economically feasible for industry to develop commercially. Private companies now provide satellite-based services and have expanded the services that can be

---

[5] Joseph N. Pelton, "The History of Satellite Communications," Chapter 1 in *Exploring the Unknown – Selected Documents in the History of the U.S. Civil Space Program, Volume II: Using Space,* John M. Logsdon set al., ed. (Washington: GPO: 1998), p.7.

offered. These advances, coupled with changing economic and political realities, and the dynamics of international relations, have all altered the environment in which Comsat, Intelsat, and Inmarsat operate. Industrial, organizational, and technological changes have led many to question the market dominance of Comsat, Inmarsat, and Intelsat, and the legal and regulatory processes that provide oversight for their U.S. operations.

Today's Structure and Regulation of Comsat

The *Satellite Act* created the Communications Satellite Corporation (Comsat) to exclusively represent the U.S. in the international satellite communications marketplace. Comsat was authorized to assist in developing a global system that would, by definition, be accessible to all nations, including those to which service would show no immediate profit. That global system became Intelsat, an intergovernmental treaty organization that owns and operates a global network of communications satellites. Comsat was incorporated in the District of Columbia in 1963, and was organized so that no single company could dominate the system through ownership or patent control. (More details on the history of the *Satellite Act,* and on Comsat's creation and early regulation, are contained in Appendix A.)

Although Comsat was created as an independent and publicly-owned company, the corporation is still regulated and has clear ties to the U.S. Government. Comsat's Board of Directors consists of six representatives from the public stockholders, six representatives of the telecommunications industry, and three presidential appointees. The public owned 50% of the corporation's stock and U.S. telecommunications companies providing international services owned the remaining 50%.

Under the provisions of the *Satellite Act,* three agencies—the FCC, the National Telecommunications and Information Administration (NTIA) of the Department of Commerce, and the Department of State's Office of Communications and Information Policy—are authorized to issue joint instructions to Comsat on a wide range of matters affecting the role of Comsat in Intelsat. The FCC serves as the primary regulatory authority for the U.S. telecommunications industry, and has direct oversight of Comsat's activities. Specifically, the Act requires that the FCC: (1) ensure effective competition in the purchase of telecommunications goods and services by Comsat; (2) regulate the allocation of facilities by Comsat to the users of the satellite system; (3) ensure that Comsat's rates are appropriate; (4) approve technical characteristics of the operating system; (5) grant authority to other carriers to provide similar services; (6) authorize Comsat's initial public offering (IPO); (7) insure or require that any additional services are in the public interest; and (8) make other rules or regulations necessary to carry out provisions of the *Satellite Act.* The NTIA, on the other hand, serves primarily as an advisor to the President for policy issues in telecommunications. In particular, the Act requires that the NTIA: (1) aid in the planning, development, and execution of the commercial communications satellite system; (2) conduct a continuous review of the system and corporation; (3) work with the Department of State to ensure compliance with the *Satellite Act;* (4) coordinate the use of the electromagnetic spectrum; and (5) serve as the liaison between the President and Comsat. Substantial control of the corporation's relations with foreign governments and carriers is left to the State Department. The Trade Compliance Center reports that these agencies have routinely been given access to signatory deliberations within Intelsat's Board of Governors, and formerly in Inmarsat's Council.[6]

Today, Comsat offers communications satellite services between the United States and other countries and, on a more limited basis, between the continental United States and offshore U.S. points.[7] Comsat delivers satellite facilities and services directly to common carriers ("jurisdictional services") as required under the *Satellite Act*. These communications satellite facilities and transmission services are provided through systems owned and operated by Intelsat and Inmarsat. Services furnished through Intelsat include broadcast and digital networking services between the U.S. and the rest of the world. The services are used by Internet service providers, multinational corporations, telecommunications carriers and U.S. and foreign governments. Comsat provides maritime, aeronautical and land-mobile communications services through the Inmarsat system. The corporation also provides satellite transmission services ("non jurisdictional services"), as well as a variety of other communications products and information services, to domestic and international, commercial and government customers through its series of subsidiaries.[8] The company provides "digital networking" to connect multinational corporations and other companies in international markets. In addition, it develops satellite communication technologies and provides technical consulting services.

Intelsat Moves toward Privatization

The *Satellite Act* required Comsat officials and appropriate U.S. government agencies to establish an international satellite communications system. Therefore, shortly after its own creation, these parties began negotiations to establish an international organization to develop and manage a global commercial communications network. In 1964, an international interim agreement signed by President Kennedy created a global system known as the International Telecommunications Satellite Consortium, with Comsat as its manager. The organization, now known as the International Telecommunications Satellite Organization (Intelsat), operates the space segment of a global telecommunications satellite system providing voice, data and video services to approximately 200 countries. In accordance with the 1973 Intelsat Operating Agreement, the capital requirements of Intelsat are contributed to by the signatories of member countries. A signatory means a party to the Operating Agreement establishing Intelsat, or the telecommunications entity designated by a party, which has signed the Operating Agreement relating to Intelsat.[9] (For more information on the history of this treaty organization, see Appendix A.)

---

[6] "Advantages to International Satellite Organizations," Chapter 10 in: Trade Compliance Center, *Addressing the Challenges of International Bribery and Fair Competition: July 1999,* First Annual Report to Congress on the OECD Antibribery Convention, p. 75, [http://www.mac.doc.gov/tcc/bribery/oecd_report!], visited August 3, 1999.

[7] Comsat Corporation, "About COMSAT," [http://www.comsat.com/company_info/comp_info.htm], visited September 10, 1999.

[8] Previously, Comsat was allowed to provide services directly to non-carrier users only in unique and exceptional circumstances. However, a 1982 FCC decision now allows Comsat to provide services directly to end customers.

[9] Intelsat, "Agreement Relating to the International Telecommunications Satellite Organization Intelsat ' Article I (Definitions)," August 20, 1971, [http://www.intelsat.int/about/agremnts/aart-.1.htm], visited January 6, 2000.

Intelsat continues to operate as a commercial cooperative within the structure outlined in the Operating Agreement, and now has 143 member countries and more than 40 investing entities.[10] These owners still contribute capital in proportion to their relative use of the system, and receive a return on their investment. Users pay a charge for all services, depending on the type, amount, and duration of the service. Any nation may use the Intelsat system, whether or not it is a member, and Intelsat maintains that most of the decisions, which the member nations make, are accomplished by consensus.[11] The organization currently owns and operates 17 satellites, which offer a range of global satellite services. These services include traditional services, such as voice, data, and video transmissions, but have also been expanded to include full-motion video, sound and graphics, telemedicine, distance education and electronic commerce.[12]

Characteristics of the membership in the organization, however, have changed somewhat since its creation in 1964. In some countries, the signatory is still the national government itself in others, it is now a private or state-owned company, or a joint venture between the government and a private company. Information from the International Telecommunications Union (ITU) shows that there has been a decrease in the number of Intelsat signatories that are owned by the government of the country in which they operate.[13] Privatization is seen primarily in the more developed countries, which own relatively large portions of Intelsat (greater than 1.0%). This is in contrast to most signatories in sub-Saharan Africa and several in South America, which own substantially smaller portions of Intelsat (usually 0.05%), and are still owned or operated by that country's government.

Responding to these and other changes in the international satellite communications industry, Intelsat has begun to privatize. In 1998, Intelsat created New Skies Satellite N.Y., a commercial spin-off that operates five geostationary satellites, collectively providing complete global coverage.[14] Intelsat states that this unanimous decision by Intelsat's members is the first step in an internal transformation process.[15] Upon New Skies' incorporation in The Netherlands in November 1998, six of Intelsat's satellites were transferred to the new private company.[16] Comsat owns approximately 16% of New Skies shares.

Also in 1998, Intelsat's Assembly of Parties (the official meeting of all 143 member governments) voted to restructure and eventually commercialize Intelsat. Intelsat's management called this decision the second phase of internal restructuring, with the objective of transforming Intelsat into a fully-commercialized company by 2001.[17] On November 1,

---

[10] Intelsat, "The Agreement and Operating Agreement Relating to the International Telecommunications Satellite Organization 'Intelsat'," August 20, 1971 (entered into force: February 12, 1973), [http://www.intelsat.int/about/agremnts/agremnts.htm], visited January 6, 2000.

[11] Intelsat, "What's a Satellite? 'Mirror in the Sky,'" [http://www.intelsat.com/about/intelsat.htm], visited September 9, 1999.

[12] Intelsat, "Intelsat Makes Worldwide Communications" *Intelsat Annual Report 1998*, [http://www.intelsat.com/aboutIannrept/98/wwide.htm] visited September 9,1999.

[13] ITU, Telecommunications Development Bureau, *General Trends in Telecommunications Reform 1998*.

[14] A sixth satellite is scheduled for launch in mid-2000. A satellite has a geostationary orbit if it is in an equatorial orbit, with an orbital period equal to the earth's rotation period. In other words, the position of a geostationary 'satellite is such that the satellite is always fixed over the same place on earth. They are located approximately 22,237 miles, or *35,579* kin, above the earth's surface.

[15] Intelsat, "Management's Vision for the Future," *Intelsat Annual Report 1998*, p. 24, [http://www.intelsat.com/about/annrept/98/future5.htm], visited September 9, 1999.

[16] Intelsat, "News Releases," [http://www.intelsat.com/news/press/press.htm], December 1, 1998, visited January 5, 2000.

[17] *Ibid.*

1999, the Assembly created a working group, composed of Intelsat member governments and shareholders, to expedite the work required to finalize the restructuring effort. This group will present final restructuring recommendations at the July 2000 Assembly meeting. Intelsat expects that these plans will enable the organization to establish the private "New Intelsat" by April 2001.[18]

Intelsat asserts that, even with the advent of New Skies, it will continue to provide services to developing countries. Intelsat's "lifeline connectivity obligation" (LCO) requires that eligible Countries continue to have capacity available on the Intelsat system for a period beyond privatization. The Assembly, in making the decision to privatize, issued a statement emphasizing the importance of Intelsat's role in providing worldwide coverage, asserting that "the privatized Intelsat shall maintain the objective of providing global connectivity."[19]

Inmarsat Privatizes

The International Maritime Satellite Organization (Inmarsat) came into existence in 1979. Inmarsat grew out of an initiative of the then International Maritime Consultative Organization (IMCO), now the International Maritime Organization (IMO), an agency of the United Nations. Responding to an international call for distress and safety applications for ships at sea, Inmarsat's original purpose was to provide satellite-based shipboard communications. Like Intelsat, Inmarsat's signatories contributed the necessary capital and shared the risk involved. Comsat's functions for Inmarsat were outlined in the 1978 amendments to the *Satellite Act*, which also specified that Comsat would serve as the U.S. signatory to Inmarsat. This codification of the Inmarsat-Comsat relationship differs from the Intelsat-Comsat relationship, which was never codified. However, similar to the Intelsat-Comsat situation, Comsat has financial interest in Inmarsat facilities. (More information on the origins of Inmarsat is contained in Appendix A.)

On April 1, 1999, two decades after it was established, Inmarsat became the first intergovernmental treaty organization to privatize and become a limited company. Inmarsat stated that this transition was made so that benefits to its owners and customers could be created, and in order for the company to more effectively compete in the global mobile communications marketplace. The privatized Inmarsat is a partnership with shareholders in 81 countries, with headquarters in London. Each member country is still represented by a signatory, usually its telecommunications provider or ministry. Comsat is a shareholder and board member in the new private company, and its 22% ownership in Inmarsat was transferred when Inmarsat became private. The organization plans a public offering within approximately two years after privatization.

Currently, Inmarsat is the world's largest provider of global mobile satellite communications for commercial, and distress and safety applications, at sea, on land and in the air. Services available through the Inmarsat satellite network include direct-dial telephone, data, facsimile, telex, electronic mail, high quality audio, compressed video and still video pictures, telephoto, slow-scan television, videoconferencing and telemedicine. The system

---

[18] Intelsat "Intelsat Members Decide to Privatize," [http://www.intelsat.coni/news/press/99-30e.htm], November 1, 1999, visited January 5, 2000.

[19] *Ibid.*

consists of nine satellites in geostationary orbits. Four of these satellites provide overlapping operational coverage of the globe (apart from the extreme polar areas); the others are used as in-orbit spares or for leased capacity.

Inmarsat is the sole provider of the satellite elements of the Global Maritime Distress and Safety System (GMDSS). The GMDSS, which came into force on February 1, 1999, is designed to reinforce maritime safety by supporting safety-related services to ships. The system allows ships in all locations the ability to transmit a distress alert to shore authorities.

Inmarsat is a shareholder in ICO Global Communications, a private spin-off formed in 1995.[20] ICO is developing a new satellite system designed to provide global service to hand-held phones. ICO's network of low, Earth-orbiting satellites (LEOs) would supply services to dual-mode handsets operating with cellular systems where available, and via satellite elsewhere. From a customer's perspective, ICO's global phone service would be comparable to satellite phone systems also being developed by Iridium and GlobalStar. ICO's system of 12 satellites, to be built by Hughes Electronics, was originally expected to be in operation in the year 2000. However, in August 1999, ICO filed for Chapter 11 protection from creditors.[21] Media sources report that a representative from ICO stated that it needed additional financing to proceed with development of its satellite network.[22] The same source reports that ICO asserts that the Chapter 11 filing "should provide ICO with the extra time needed to reorganize, recapitalize and complete our financing." Hughes Electronics Corporation has stated that the company could lose about $300 million if ICO defaulted on its debts.[23] As an investor, Comsat has about $70 million invested in ICO's global system. In November 1999, Craig O. McCaw, an initial investor in the cellular telephone industry, agreed to raise $1.2 billion to bring ICO out of bankruptcy.[24] In December 1999, ICO's shareholders elected 13 new board directors.[25] ICO has been licensed by the FCC, and is still reportedly seeking access to foreign countries' markets and the licenses necessary to provide global services. The General Accounting Office (GAO) reports that potential competitors are concerned that Inmarsat's close relationship to JCO could hinder the development of competition.[26]

Increased Competition, Remaining Barriers

Both international satellite organizations now face competition from other satellite systems. Inmarsat is now competing in the land-based mobile communications market

---

[20] Information supplied to CRS by ICO Global Communications on January 5, 2000, indicates that Inmarsat currently owns approximately 8% of ICO's shares.

[21] Two weeks earlier, Iridium L.L.C., which is offering a similar service, failed to attract nearly as many customers as it needed and also filed for bankruptcy, forcing its main backer, Motorola, to write off $2.3 billion.

[22] Leslie Cauley, "ICQ Global Satellite-Phone Venture, Affiliates File for Bankruptcy Protection," *DJ via Dow Jones,* September 7, 1999.

[23] Sam Silverstein and Warren Ferster, "ICO Goes Bankrupt: Hughes Has $300 Million at Stake," *Space News,* vol. 10, no. 33, September 6, 1999, p. 1.

[24] Andrew Pollack, "Out on a Limb as Technologies Converge: True Believers Push Satellite Telephones," *New York Times,* January 3, 2000, p. 3.

[25] ICO Global Communications, "ICO Shareholders Elect Board Directors," December 20, 1999, [http://www.ico.com/press/index.htm], visited January 4, 2000.

[26] GAO, *Telecommunications: Competitive Impact of Restructuring the International Satellite Organizations,* (Washington: GPO, July 1996), p. 8.

through ICO, its private spinoff company. At the same time, Intelsat faces more competition in providing jurisdictional services.[27] Some of Intelsat's competitors have established themselves as strong players in the international satellite communications marketplace. For example, one competitor, the Pan-American Satellite Corporation (PanAmSat), now possesses 20 satellites and seven technical ground facilities.[28] Another, Loral CyberStar (formerly known as Orion), provides telecommunications services to approximately 5000 sites in about seventy countries.[29]

Despite this increased competition, the Trade Compliance Center (TCC) reports that some U.S. firms that compete with Intelsat and Inmarsat, such as PanAmSat and Orion (now CyberStar), face significant barriers to providing domestic and international service.[30] Furthermore, as these companies still provide predominately video services, their potential customers remain somewhat separate from those of Intelsat, which provides video, voice, data, and Internet services. According to the TCC, this segregation between voice and video markets remains a residual source of concern. Moreover, U.S. satellite communications companies seeking to compete with Intelsat or Comsat often have difficulties in obtaining operating agreements with foreign administrators.[31] The TCC reports that although some of these barriers are coming down, they are still a serious problem.

Competition in the video market has increased somewhat in the last two decades, largely due to efforts made by Congress and the FCC in promoting increased competition within the U.S. telecommunications industry.[32] However, the GAO concurs with the TCC and reports that, as separate U.S. satellite systems have been restricted from providing all international telephone services, competition has emerged primarily in the form of fiber optic cables.[33]

The U.S. situation contrasts with many other nations where telecommunications facilities either are government-owned, -operated, or -controlled. Frequently, in these situations the

---

[27] This competition between service providers resulted primarily from a 1985 decision by the FCC to allow separate U.S. -owned systems to compete with Intelsat. One of these separate systems, PanAmSat, launched its first satellite over the Atlantic Ocean Region in 1988, becoming the first private-sector international satellite service provider. Orion, a similar example, was founded in 1982 and began construction oil its first satellite in 1991. Orion 1 was launched in 1994, and offered Trans-Atlantic and Pan-European communications to multinational corporations. In 1998, Orion was acquired by Loral Space & Communications, becoming Loral CyberStar.

[28] PanAmSat, "The Satellites: The PanAmSat System," available on the Internet at [http://www.pananisat.com/sat/system.htm], visited December 29, 1999.

[29] Loral CyberStar, "CyberStar's Global Reach," also available on the Internet at [http://209.239.66.33/international/index.jsp], visited December 29, 1999.

[30] "Advantages to International Satellite Organizations," p. 75.

[31] According to the TCC, an American-owned satellite system that seeks to provide telecommunications services between the U.S. and the public network of a foreign country must obtain an operating agreement with that country. Whether or not such operating agreements will be allowed is a matter of national telecommunications policy. Any country that seeks to protect Intelsat or Inmarsat can refuse to authorize a competitor to operate there. For example, in its filing with the NTIA, PanAmSat noted that, while it can provide full-time video service in 129 countries, only eight countries allow PanAmSat to offer switched voice traffic. The company is also not currently allowed to offer any voice service in five global regions: Western Europe, Eastern Europe, North America, Central and South Asia, and the Middle East.

[32] U.S. Congress, House Committee on Science and Technology, Subcommittee on Space Science and Applications, "Communications Satellites," Chapter 3 in *United States Civilian Space Programs, Volume II: Applications Satellites,* committee print prepared by the Congressional Research Service, Library of Congress, 98th Congress, 1st Session (Washington: GPO, 1983), p. 42.

[33] GAO, *Telecommunications: Competition Issues in International Satellite Communications,* (Washington: GPO, October 1996), p. 4. In the market for certain types of international television and video services, such as broadcasting, GAO reports that U.S. satellite companies and systems compete alongside other international, regional and domestic systems, and have become viable competitors.

government-owned PTTs are also the regulatory authorities, which decide which satellite systems will have access their domestic markets, and this often limits the amount and type of competition allowed. However, in recent years, the pace of change in the international telecommunications sector has been both rapid and dynamic. In several countries the incumbent telecommunications operators—such as PTTs—have been replaced or augmented by privately-owned carriers. In other countries, formerly insulated domestic markets have been opened to the entry of new operators, introducing competition and often driving down prices of telecommunications services. The ITU reports that, for many developing countries, privatization is currently the most important item on the agenda for transforming the telecommunications sector.[34] Comsat, agreeing with the ITU's analysis, stated in its comments to the NTIA:

> today about 75% of Intelsat's ownership is held by companies that are fully or partly privatized. In fact, of those signatories with an ownership share of 0.5% or more, all but six are fully or partly private, and four of those six have announced plans to privatize in the near future.[35]

However, the ITU also reports that, in the long run, opening markets to competition "will almost certainly be more significant and profound" than privatization.[36] For this reason, more and more countries, including those with membership in Intelsat and Inmarsat, are struggling with complicated developments when a government pursues a policy of both privatizing and liberalizing the market to allow for the entry of new carriers. The result is a dynamic international telecommunications sector, with continual changes in the policies of other governments and almost constant new introductions of private carriers throughout the world.

LEGISLATIVE ISSUES

During the 106th Congress, Congress considered legislation related to the process by which the international satellite organizations will privatize. While these organizations have made initial moves toward privatization, issues remained about the speed with which privatization occurs, the rules by which access to certain markets is authorized, and the final organizational structure of each privatized satellite organization. Three bills seeking to amend the *Satellite Act* were introduced in the 106th Congress: H.R. 1872 (Bliley), the *Communications Satellite Competition and Privatization Act of 1998;* S. 1328 (Inouye), the *Communications Satellite Competition and Privatization Act of 1997;* and S. 2365 (Burns), the *International Satellite Communications Reform Act of 1998.* Only H.R. 1872 passed the House, and none passed the Senate.

During the first session of the 106th Congress, two related bills were introduced: S. 376 (Burns), the *Open-market Reorganization for the Betterment of International Telecommunications (ORBIT,) Act;* and H.R. 3261 (Bliley), the *Communications Satellite*

---

[34] ITU, Telecommunications Development Bureau, *General Trends in Telecommunications Reform 1998,* Volumes II —VI (Geneva: June 1998).

[35] "Advantages to International Satellite Organization." P. 76.

[36] ITU, Telecommunications Development Bureau, *General Trends in Telecommunications Reform 1998,* Volumes I (Geneva: June 1998), p. 3.

Competition and Privatization Act of 1999. S. 376 passed the Senate. In the House, S. 376 passed, with the text of H.R. 3261 as an amendment in the nature of a substitute. The most notable differences between the House and Senate versions included "direct access" provisions, differences in privatization deadlines, and criteria for authorizing access to the U.S. market. Both versions provided incentives for Intelsat and Inmarsat to more quickly and completely privatize, and required Comsat to compete in a comparatively more open satellite services market.

In support of this congressional activity, some Members of Congress contended that Intelsat and Inmarsat enjoy privileges and immunities that give them unfair competitive advantages over other companies.[37] These parties wanted to ensure that the organizations privatize in a "pro-competitive" manner by using access to the U.S. market and Comsat's role as U.S. signatory as leverage.[38] Other Members, while agreeing that Intelsat and Inmarsat should privatize, disagreed with legislation that might try to unfairly hold Comsat responsible for that process, and potentially cause it financial hardship.[39] They also stated that it would be inappropriate for the U.S. to force its policies on international treaty organizations, which are also responsible to many other signatory nations.[40]

Explained below are significant issues discussed by Congress while considering this legislation.

Lockheed Martin Plans to Buy Comsat

In September 1998, Lockheed Martin announced its plans to acquire Comsat. On August 20, 1999, 99% of Comsat's shareholders voted in favor of the merger.[41] At that time, company ownership was restricted by FCC and Department of Justice (Justice) rules and regulations, as well as the *Satellite Act.* However, on September 15, 1999, the FCC announced its approval of Lockheed's "common carrier" status, and thus, approved its purchase of 49% of Comsat's stock. Then, on September 16, 1999, Justice announced its approval of Lockheed's plans to purchase 49% of Comsat.[42] Therefore, after both regulatory hurdles were overcome, Lockheed Martin purchased 49% of Comsat's shares for $45.50 per share.[43]

---

[37] Rep. Tom Bliley, remarks in the House, *Congressional Record,* daily edition, vol. 144, no. 55, May 6, 1998, p. H2824-H2825.

[38] Rep. Anna Eshoo, remarks in the House, *Congressional Record,* daily edition, vol. 144, no.55, May 6,1998, p. H2827.

[39] Rep. Paul P. Gillmor, remarks in the House, *Congressional Record,* daily edition, vol. 144, no. *55,* May 6, 1998, p. H2828.

[40] Rep. John D. Dingell, remarks in the House, *Congressional Record,* daily edition, vol. 144, no. 55, May 6, 1998, p. H2826.

[41] Sam Silverstein, "Comsat Shareholders Vote In Favor Of Merger With Lockheed Martin," *Space News,* September 6, 1999, p. 14.

[42] The Justice Department (Justice) had to agree that antitrust laws will not be violated in the completion of Lockheed Martin's purchase of Comsat. It had been expected that Justice would demand that, before buying Comsat, Lockheed must sell its stock in Loral Space & Communications Ltd., which is also in the satellite communications business. Contrary to expectations, the department did not require Lockheed to divest its stake in Loral.

[43] Comsat Corporations, "Department of Justice Clears Comsat-Lockheed Martin Merger," September 16, 1999, [http://www.comsat.com/news/news_set.htm], visited September 16, 1999.

Lockheed Martin stated that it planned to eventually acquire the remaining 51% of Comsat in a one-for-one stock swap.[44] Comsat would then be absorbed into its new parent company. However, in order to prevent any company or group of companies from controlling the ownership of Comsat, the *Satellite Act* sets private ownership limits at 49% for Comsat. For Lockheed to absorb the remaining shares of Comsat, Congress would have to legislate a change in the ownership cap for Comsat.

"Fresh Look," "Takings," and the *Tucker Act*

Given increased competition in the satellite communications industry, and the emergence of new satellite system providers such as PanAmSat and Loral Orion, some Members of Congress wanted Comsat's contracted customers to have an opportunity to reexamine those contracts.[45] This legislative provision would have allowed a "fresh look" time period during which users or providers of telecommunications services could renegotiate the terms, rates, conditions, or other provisions of previously entered contracts or tariff commitments with Comsat. Supporters asserted that the "fresh look" provision was a tool intended to accelerate the transition to open competition.[46]

Several Members of Congress also wanted to require both Intelsat and Inmarsat to privatize certain activities at specific levels before the organizations would be allowed to either directly access the U.S. market, or provide additional services beyond what is currently provided through Comsat. They stated that such incentives would serve to expedite "pro-competitive" privatization efforts by the organizations, and ultimately the end consumer.[47]

However, other Members of Congress voiced concerns with both provisions. They maintained that a renegotiation of contracts would constitute an unconstitutional "taking" of property and a "violation of the *Tucker Act*."[48] These individuals opposed additional requirements for Intelsat and Inmarsat, stating that they are too strict and rigid, and would result in closing the market to both organizations' spinoffs and successors.[49]

The Fifth Amendment of the Constitution prohibits the "taking" of private property for public use without just compensation. This constitutional Takings Clause is self-executing. The *Tucker Act* is the principle jurisdictional statute for the U.S. Court of Federal Claims (CFC).[50] The functions of the CFC are: (1) waiving the sovereign immunity of the United States to certain types of money claims, and (2) providing jurisdiction over such claims when they arise. Thus, when saying that a government action "constitutes a 'taking'" and "violates the *Tucker Act*," legal experts usually interpret this to be a claim that the government has

---

[44] "FCC OKs Lockheed Stake in Comsat," *Reuters*, September 15, 1999, [http://www.washingtonpost.com/ wp-srv/business/daily/sept99/lockheed15.htm], visited September 15, 1999.

[45] Rep. Christopher Shays, remarks in the House, *Congressional Record,* vol. 144, no. 55, May 6, 1998, pp. H2830-H2831.

[46] Rep. Rodney P. Frelinghuysen, remarks in the House, *Congressional Record*, vol. 144, no. 55, May 6, 1998, p. H2829.

[47] Rep. Tom Bliley, remarks in the House, *Congressional Record*, vol. 144, no. 55, May 6, 1998, p. H2825.

[48] Rep. John D. Dinell, remarks in the House, *Congressional Record*, vol. 144, no. 55, May 6, 1998, p. H2826.

[49] Rep. Constance A. Morella, remarks in the House, *Congressional Record*, vol. 144, no. 55, May 6, 1998, p. H2840-H2841.

[50] 28 U.S.C. §1491.

taken property without just compensation, and that this claim against the government is assertable in the CFC via the *Tucker Act*.[51]

Supporters of the "fresh look" provision and entry requirements for Intelsat, Inmarsat, its successors, or spinoffs, argued that these provisions would not constitute a "talking." These individuals maintained that the provisions would not take Comsat's property in its contract rights, but only give customers the right to renegotiate.[52] Furthermore, these Members of Congress asserted that Congress reserved the right to regulate satellites and change these legal understandings at any time since Congress created Comsat through the *Satellite Act,* and codified the relationship between Comsat and Inmarsat in amendments to that Act.[53]

Opponents of the "fresh look" provision argued that Congress has no right to impair the obligations of these contracts, and that this action would be equivalent to an unconstitutional "taking" of property.[54] These individuals also asserted that the setting of strict privatization requirements for Intelsat and Inmarsat constitutes a breach of contract between the United States and Comsat, which they maintain is also a "taking" of property.[55]

"Direct Access"

Until September 1999, Comsat was the only U.S. company that was authorized to sell communications services through the Intelsat or Inmarsat system, acting as a middleman in providing access to these global satellite systems for the rest of the U.S. communications industry. Some U.S. telecommunications companies contend that this prohibition of "direct access" to the Intelsat and Inmarsat constellations results in increased costs to end customers. In support of this argument, one media source reports that, according to the FCC, Comsat obtains an average markup of 68% on the services provided by Intelsat.[56]

For these reasons, a key issue of debate within Congress was the inclusion of a "direct access" provision in privatization legislation. This provision would allow either international satellite organization to sell its services directly to U.S. telecommunications companies, without using Comsat as a middleman in these transactions. Members of Congress in support of a "direct access" provision maintained that the current lack of access, except through Comsat, distorts the U.S. market and results in inflated prices for consumers.[57] On the other side of the issue, those opposed to this provision asserted that, under the *Tucker Act* and the Takings Clause of the Constitution, the U.S. government would be liable to Comsat for lost business as a result of allowing "direct access."

The *Satellite Act* gives the FCC the authority to make the rules and regulations necessary to carry out the Act's provisions in the public interest. Utilizing the Act's authority, on September 16, 1999, in addition to approving Comsat's acquisition by Lockheed, the FCC stripped Comsat of its exclusive right to sell transmission space over the Intelsat system.

---

[51] Analysis provided by the American Law Division of CRS.
[52] Rep. Tom Bliley, remarks in the House, *Congressional Record*, vol. 144, no. 55, May 6, 1998, p. H2832.
[53] *Ibid.*
[54] Rep. John D. Dingell, remarks in the House, *Congressional Record,* vol. 144, no. 55, may 6, 1998, p. H2840.
[55] Rep. Constance A. Morella, remarks in the House, *Congressional Record,* vol. 144, no. 55, May 6, 1998, p. H2840.
[56] Sheila Foote, "House Passes Intelsat Privatization Act," *Defense Daily,* November 16, 1999, p. 3.
[57] Rep. Jim Davis, remarks in the House, *Congressional Record*, vol 144, no. 55, May 6, 1998, p. 2835.

Shortly thereafter, Comsat filed a petition to stay the FCC's decision in U.S. court.[58] The court's decision notwithstanding, interexchange carriers, broadcasters, and earth station operators are now able to bypass Comsat in the acquisition of Intelsat services. Under the FCC's September 1999 ruling, any company can purchase services directly from Intelsat as a "direct access" customer. Comsat will collect a signatory fee tariff of 5.58% on entities that bypass the corporation in directly accessing the Intelsat or Inmarsat constellation.[59] The FCC's ruling and the tariff went into effect on December 6, 1999.

Under Intelsat's "direct access" guidelines, a signatory can authorize access by certain non-signatory entities operating within its territory ("appointed customers") to Intelsat at any of four access levels.[60] In general, the users are required to invest in proportion to the relative usage of the system, and at the highest access level (Level 4), users can invest directly in the organization. The September 1999 ruling by the FCC allows "direct access" up to Level 3. At least one U.S. non-signatory has already taken advantage of the change in "direct access" policy. In early January 2000, media sources stated that Spacelink International, a Washington, D.C.-based reseller of international satellite capacity, reportedly obtained direct-access rights to the Intelsat fleet of telecommunications satellites.[61]

Some Members of Congress objected to the FCC's ruling on this issue. They maintained that simply authorizing "direct access" to occur is not sufficient. Instead, these Members of Congress wanted to use the pending acquisition of Comsat by Lockheed Martin as leverage for encouraging full "direct access" to Intelsat. Under a legislative provision (H.R. 3261, Sec. 645) offered by these Members, before ownership caps on Comsat would be lifted, allowing the purchase of the remaining 51% of the corporation by Lockheed Martin, Intelsat would have had to achieve Level 4 "direct access" in the U.S. telecommunications market. The legislative provision directed the FCC to "take such steps as may be necessary" to allow this "direct access" to take place.

ACTIVITY IN THE 105TH CONGRESS

Three bills (S. 2365, H.R. 1872, and S. 1328) were introduced during the 105th Congress. Only one passed the House, and none were enacted into law. As these bills are similar to those currently pending in the 106th Congress, below is a brief account of the legislative history and provisions of legislation in the 105th Congress.

---

[58] In October 1999, Comsat filed a lawsuit in the U.S. Court of Appeals seeking to stay implementation of the FCC's order (ease no. 99-14 12). Two weeks later, the FCC denied Comsat's earlier request for the agency to stay its own order. Before denying Comsat's stay request, several telecommunications companies—including AT&T Corporation, MCI WorldCom, Inc., and Sprint Corporation—filed comments with the FCC in opposition to Comsat's request. These petitions remain pending. For more information, see: "FCC Sets Dec. 6 As Date for Intelsat Direct Access," *Telecommunications Reports,* October 18, 1999, p. 26.

[59] In its September ruling, the FCC suggested a *5.58%,* but left the issue somewhat vague, saying that Comsat could increase the fee with sufficient justification. Though many observers had reportedly expected Comsat to file for a higher fee, in early December 1999, Comsat filed with the FCC a signatory fee tariff of just *5.58%.* For more information, see: Terry Banks, "Comsat Will See 5.58% Direct Access Signatory Fee," *Communications Daily,* December 2, 1999, p. 3.

[60] Level 1 is operational and technical access. Level 2 includes commercial and service matter access. Level 3 is for contractual access, including direct ordering of and liability for paying for Intelsat capacity. Finally, Level 4 is utilized for investment access, including direct investment in Intelsat. Further information on these levels of "direct access" to Intelsat may be obtained at their Internet website: [http://www.intelsat.int].

S. 2365: The Burns Bill

During the second session of the 105th Congress, Senator Conrad Burns sponsored S. 2365, the *International Satellite Communications Reform Act of 1998*. This legislation sought to amend the *Satellite Act* so as to promote a competitive global market for satellite communications services, and encourage the continued restructuring, on a privatized basis, of Intelsat and Inmarsat.

Most notably, the bill would have directed the Secretary of State and all other agencies to pursue specific privatization objectives and open commercial competition through international negotiation within established organizations and directly with other interested nations. If these objectives were not achieved by January 1, 2003, the bill required that the President take appropriate implementation steps. Specifically, the bill would have directed Comsat, as the signatory to both international satellite organizations, to take all necessary steps to: (1) restructure both organizations so that private competition increases; (2) ensure that Intelsat's restructuring plan is implemented in a pro-competitive manner; and (3) establish safeguards to ensure the establishment of pro-competitive successors and affiliates of both organizations. In addition, the bill would have required the FCC to utilize specific privatization standards in granting authority to enter the U.S. market to any satellite communications entity. Under this legislation, Comsat would have lost some privileges and immunities it enjoys as U.S. signatory to Intelsat and Inmarsat.

The Communications Subcommittee of the Senate Commerce Committee held hearings on the bill (S.Hrg. 105-829), and the full committee marked it up. However, the legislation did not make it to the Senate floor before the close of the 105th Congress.

H.R. 1872: The Bliley Bill

Also during the second session of the 105th Congress, Representative Tom Bliley introduced H.R. 1872, the *Communications Satellite Competition and Privatization Act of 1998*. Senator Daniel K. Inouye introduced an identical bill (S. 1328) in the Senate.

This legislation sought to amend the *Satellite Act* so as to encourage the privatization of Intelsat and Inmarsat, and to promote robust competition within the international satellite communications industry. Most notably, this bill would have required the President to seek full privatization of Intelsat by January 1, 2002, and of Inmarsat by January 1, 2001. Failure to do so would have precluded any entity subject to U.S. jurisdiction from accessing the Intelsat or Inmarsat constellation to provide services to, from, or within the U.S. after that date. Included were provisions, which would have required both the FCC and the President to ensure that privatization is implemented in a pro-competitive manner. The bill also would have required that, with FCC approval, both Intelsat and Inmarsat have "direct access" to the U.S. market as of January 1,2000, and direct investment (Level 4 "direct access") in Intelsat by U.S. companies as of January 1, 2002.

This legislation would have provided a "fresh look" period for Comsat's contracted customers, during which companies with existing multi year agreements with Comsat could renegotiate these contracts. Under this legislation, Comsat would have lost all privileges and

---

[61] "Spacelink Receives Rights to Intelsat Satellite Fleet," *Space News,* January 10, 2000, p. 16.

immunities it enjoys as U.S. signatory to Intelsat and Inmarsat. Finally, the bill would have allowed Comsat's acquisition.

Following much debate, the bill passed the House (403-16) on May 6, 1998. However, the Senate failed to act on the legislation during the 105th Congress.

106TH CONGRESS ACTIONS

S. 376: Senator Burns' *ORBIT* Bill

The debate resumed during the first session of the 106th Congress, when S. 376, the *Open-market Reorganization for the Betterment of International Telecommunications (ORBIT) Act,* was introduced by Senator Conrad Burns. This bill seeks to amend the *Satellite Act* so as to promote a fully competitive satellite communications domestic and international market by encouraging the privatization of Intelsat and Inmarsat, and by reforming the regulatory framework of Comsat.

Similar to Senator Burns' legislation in the 105th Congress, the bill defines the policy of the U.S. to be: (1) encouraging pro-competitive privatization of Intelsat as soon as possible, and no later than January 1, 2002; (2) working with Intelsat and its international partners to achieve this goal; and, (3) encouraging Inmarsat to implement its privatization fully. Most notably, the bill allows Comsat's acquisition, requires the President to seek full privatization of Intelsat and Inmarsat by January 1, 2002, and directs the FCC to ensure that privatization is implemented in a pro-competitive manner, as defined by specific criteria. Under this bill, the President is required to report to Congress on the extent of privatization and on the potential for U.S. market distortion following Intelsat's final privatization. When considering applications for U.S. market entry, the bill requires that the FCC determine whether potential anti competitive or market-distorting consequences exist in any continued relationship between Inmarsat or Intelsat and their affiliates or successors, such as ICO or New Skies. The bill requires that Comsat lose some privileges or immunities it enjoys as signatory to Intelsat and Inmarsat.

A controversial provision is the prohibition of either organization's "direct access" to the U.S. market until July 1, 2001, or sooner if the FCC determines that the organization has achieved a pro-competitive privatization status. Also prohibited in the bill is any "fresh look" time period. The Senate passed S. 376 on July 1, 1999, by unanimous consent.

H.R. 3261: The Bliley Bill

Also during the first session of the 106w Congress, Representative Tom Bliley, Chairman of the House Commerce Committee, introduced the *Communications Satellite Competition and Privatization Act of 1999,* H.R. 3261.

Similar to Representative Bliley's bill during the 105k" Congress (H.R. 1872), H.R. 3261 seeks to amend the *Satellite Act* so as to promote a fully competitive global market for satellite communication services by fully privatizing the intergovernmental satellite organizations, Intelsat and Inmarsat. Most notably, this bill requires the President to seek full privatization of Intelsat by April 1, 2001, and of Inmarsat by April 1, 2000. (These dates are eight months earlier than those deadlines defined within H.R. 1872 in the 105th Congress.) Failure to meet this deadline would preclude either Intelsat or Inmarsat, and by extension, Comsat, from expanding into additional services in the U.S. market.[62] In addition, the bill requires each organization to have significantly dispersed its ownership via an IPO of its stock by the same deadline. Also included were provisions, which require both the FCC and the President to ensure that privatization is implemented in a pro-competitive manner. Before granting or renewing licenses or permits for the provision of satellite communications services within the U.S., the bill requires that the FCC determine whether such authorization will harm competition in the telecommunications market of the United States. Comsat would lose all privileges and immunities it enjoys as U.S. signatory to Intelsat and Inmarsat.

The bill would require that, with FCC approval, both Intelsat and Inmarsat have "direct access" to the U.S. market as of April 1, 2000, and direct investment (Level 4 "direct access") in Intelsat by U.S. companies as of April 1, 2001. (Similar to the bill's deadlines for privatization, these dates are eight months earlier than those set in H.R. 1872.) Upon implementation of Level 4 "direct access" to Intelsat, the bill would allow Comsat's acquisition. However, unlike the bill's predecessor, this bill does not provide a "fresh look" period. A media source reports that, in an effort to compromise, this provision was dropped because Representative Bliley wanted a "'bulletproof bill" to which it would be difficult to raise 'reasonable objections."[63]

On November 10, 1999, the day after H.R. 3261 was introduced, the House passed S. 376 by a voice vote. The text of H.R. 3261 was included as an amendment to S. 376 in the nature of a substitute.

The Corporate Positions

The corporate structures affected by satellite privatization legislation have offered varying views on the pending bills. Comsat stated that legislation which "attempt[ed] to dictate privatization" was "unrealistic."[64] On the other hand, Comsat was eager to be fully acquired by Lockheed Martin, which would require congressional action to lift the 49% ownership cap. Therefore, Comsat strongly criticized Representative Bliley's legislation (House-passed S. 376, containing the text of LLR. 3261). A Comsat spokesperson was quoted as claiming that the House bill would effectively cap Comsat's revenues at existing levels.[65] Therefore, the corporation supported Senator Burns' bill (Senate-passed S. 376, the *ORBIT* bill), which it contended was less strict in defining a specific privatization path. Also important to Comsat was the delay in "direct access," and thus, direct competition to Comsat, included in Senator Burns' bill. The Senate-passed bill allowed "direct access" only after Intelsat achieves pro-competitive privatization status, as defined in the bill. A spokesperson

---

[62] The Bliley bill defines these "additional services" as Internet, high-speed data, interactive, DTH/DBS, and Ka-band.

[63] "Ownership Caps Would Be Lifted: Bliley Offers Intelsat Privatization Bill Minus 'Fresh Look'," *Communications Daily*, November 10, 1999, p. 5.

[64] Comsat Corporation, "Inmarsat privatization progresses rapidly, Comsat Pleased That Inmarsat Assembly Endorses Plan for Full Privatization," April 27, 1998, [http://www.comsat.com/mcs/over_set.htm], visited September 10, 1999.

from Comsat asserted that the *ORBIT* bill would "establish fair and pro-competitive deregulation of the international satellite telecommunications industry."[66]

Lockheed Martin, which sought to acquire the remainder of Comsat, supported Comsat's position and disagreed with the "direct access" language in the House-passed S. 376. Lockheed Martin argued that the bill's Level 4 "direct access" to communication satellites would have "adversely affect[ed] the economic rationale" of completing its acquisition of Comsat.[67] Lockheed Martin's spokesperson is quoted as saying that this provision "would [have] force[d] Comsat to divest its investment in Intelsat." Another news article reported that a Lockheed Martin spokesperson stated that Level 4 "direct access" might prevent the corporation's final acquisition of Comsat, and that legislation including this provision "could very well stop the deal in its tracks."[68]

Intelsat also criticized the House-passed S. 376, and called the bill's passage "unfortunate" and "a solution looking for a problem."[69]

Instead, the organization asserted that it already planned to be fully privatized, following its own goals and methods, by the first quarter in 2001. A spokesperson from the organization stated that it was already committed to achieving privatization as quickly as possible for business reasons, and that this process "will not occur unless it is achieved through multilateral negotiation and by consensus of the 143 member nations that comprise Intelsat."[70]

Intelsat argued that this privatization process cannot be achieved though legislative action by the U.S. Congress, "regardless of how well-intentioned that legislative effort might be."

Intelsat was particularly concerned about provisions in the House-passed bill regarding the reallocation of orbital slot locations.[71] Intelsat asserted that this provision could "gravely compromise" Intelsat's future plans, and its abilities to "compete for emerging applications requiring technologically-advanced satellites.[72] Concerning the IPO deadline, Intelsat argued that this deadline would force Intelsat into an IPO "before it could prove its commercial potential under the enormous cloud created by H.R. 3261's [orbital] slot restrictions." Intelsat pointed out that the House-passed bill would also require Intelsat to disenfranchise former signatories that still control access to their own national telecommunication markets. Intelsat claimed that these limitations would prevent "many of Intelsat's developing country members from exercising their shareholder rights" to participate in Intelsat activities, and preclude Intelsat from competing with other communications entities, such as Iridium. Finally, Intelsat disagreed with provisions of the House-passed bill that would have prevented Intelsat from providing additional services until it met specific privatization criteria, defined in the bill. The

---

[65] Sheila Foote, "Intelsat Privatization Bill is Introduced in House," *Defense Daily*, November 12, 1999, p. 4.

[66] Comsat Corporation, "Comsat CEO Betty C. Alewine Applauds Senate Approval of S. 376," July 2, 1999, [http://www.comsat.com/news/], visited January 5, 2000.

[67] Sheila Foote, "Intelsat Privatization Bill is Introduced in House."

[68] "FOCUS: Lockheed Says Congress Could Nix Comsat Deal," *Reuters,* January 27, 2000.

[69] Intelsat, "Intelsat DG & CEO Comments on House Passage of H.R. 3261," November 10, 1999, [http://www.intelsat.com/newslpress/99-31e.htm], visited January 5, 2000.

[70] Intelsat, "Intelsat Statement on Satellite Reform Legislation," February 10, 1999, [http://www.intelsat.com/news/press/99-02.htm], visited January 5, 2000.

[71] The House-passed bill (H.R. 3261) would require Intelsat and Inmarsat to forfeit orbital slots that were neither being used as of March 26, 1998, nor designated for future use as of the same date. These forfeited orbital slots would be returned to the ITU for reallocation.

[72] Intelsat, "Update on U.S. Legislative Issues," November 17, 1999, [http://www.intelsat.com/news/policy/pletter17nov.htm], visited February 1, 2000.

organization called these criteria "stringent," and asserted that the provision will "clearly inhibit consumer choice."[73]

A representative from Inmarsat, which is already privatized, stated that the corporation was extremely concerned with the House-passed bill, and that the bill was "based on incorrect assumptions about Inmarsat, and its effects, if passed into law, w[ould] seriously affect Inmarsat operations generally and the Initial Public Offering."[74] Inmarsat was restructured as a U.K. company in April 1999, and plans an IPO by the second quarter of 2001. As stated earlier, Intelsat's headquarters are currently located in Washington, D.C. However, company officials reportedly threatened to establish the organization's headquarters outside the United States when its privatization is completed due to concerns about harsh treatment from Congress.[75]

Similar to Intelsat, Inmarsat was particularly concerned about the bill's requirements for the reallocation of orbital slots, as well as provisions regarding future forfeiture of spectrum.[76] Furthermore, the corporation asserted that the House-passed bill would "unfairly penalize Inmarsat for operating in markets that are not fully competitive," and that the company "has no control over the operation of national markets, and should not be punished for anti competitive policies over which it has no control."[77] Finally, Inmarsat disagreed with a provision of the House-passed bill that prohibited linkages between Inmarsat and ICO for 15 years. Inmarsat stated that, after the completion of the transfer of ICO's control to an investment group headed by Craig McCaw, Inmarsat would own less than 1% of TCO. Inmarsat asserted that this prohibition was unreasonable as "there should be no restrictions against future linkages in the volatile and risky mobile satellite services market."[78]

On the other side of the issue, PanAmSat endorsed the House-passed S. 376, which it contended provides open market access and competition in international telecommunications. The company asserted that privileges and immunities afforded to Intelsat and Comsat, as "treaty-protected organizations," are "potentially anti competitive and harmful to commercial satellite service providers."[79] Furthermore, a spokesperson from PanAmSat stated that it is "only fair that Congress allow Intelsat users to also be investors in the system."[80] Another spokesperson from PanAmSat commended Representative Bliley "for his long fight to create a business environment that will place Intelsat and its private competitors on a more equal footing.[81]

---

[73] *Ibid.*

[74] Inmarsat policy position paper provided to CRS by Inmarsat, January 7, 2000.

[75] "FCC International Bureau Chief Sees U.S. Regulatory Arena as the Best for Intelsat," *Satellite News* , April 24, 2000, p. 1.

[76] In addition to provisions for orbital slot reallocation, the House-passed bill (H.R. 3261) would require that Inmarsat forfeit all operating spectrum on January 1, 2006, or at the end of the useful life of existing satellites, whichever occurs later. The forfeited spectrum would then be made available for assignment to all systems, including Inmarsat, on "a nondiscriminatory basis and in a manner in which continued availability of the GMDSS is provided." The bill prohibits this spectrum from being transferred between Inmarsat and ICO.

[77] Inmarsat policy position paper provided to CRS by Inmarsat, January 11, 2000.

[78] *Ibid.*

[79] PanAmSat Corporation, "PanAmSat Applauds Introduction of Legislation to Promote Competition and Privatization in Satellite Communications," June 12, 1997, [http://wwwpanamsat.conilmedia/ pressview.asp?article=109], visited January 5, 2000.

[80] "Bliley's Bill Predates Merger," *Dow Jones Newswires via Dow Jones*, January 27, 2000.

[81] PanAmSat Corporation, "PanAmSat Applauds Congressional Approval of Legislation That Eliminates Intelsat and Comsat's Privileges and Immunities in Commercial Activities," October 22, 1998, [http://www.panatnsat.com/media/pressview.asp?article=1029], visited January 5, 2000.

No public statements was made about the pending privatization legislation by Loral CyberStar (formerly Loral Orion), ICO Global Communications, or New Skies Satellites, N.Y.

S. 376 in Conference

Before conference, several controversial differences remained between the two bills. Most notably, the House-passed S. 376 set stricter standards for deregulating Intelsat, and was more favorable to startup satellite companies, such as PanAmSat and Loral CyberStar. Moreover, the House bill allowed U.S. telecommunications companies to more quickly directly access Intelsat, without using Comsat as a middleman. "Direct access" was specifically prohibited until July 1, 2001, in the Senate bill. The Senate-passed bill allowed "direct access" sooner than this date only if the FCC determined that the organization had achieved pro-competitive privatization status.

Actions by the FCC, Justice, and each of the international organizations, which achieved some of the goals of the House version, made resolving differences in conference more complex. Additional changes in both Intelsat and Inmarsat's organizational structures also affected the final version of the legislation. In the Senate, Senators McCain, Stevens, Burns, Hollings, and Inouye were appointed to the conference committee. Speaker of the House Hastert chose Representatives Bliley, Tauzin, Oxley, Dingell, and Markey as conferees from the House.

On February 17, 2000, the news media reported that discussion among Representative Bliley, Representative Edward J. Markey, and Senator Burns led to a legislative compromise. These Members of Congress reportedly agreed to draft a bill for review by the conferees that: (1) requires the "pro-competitive privatization of Intelsat and Inmarsat"; (2) "ends Comsat's monopoly over Intelsat" by allowing U.S. companies to directly access the Intelsat constellation; (3) "ends privileges and immunities from the law" for Intelsat and Comsat; and, (4) allows the acquisition of Comsat by Lockheed Martin by "eliminating the [49%] ownership cap on Comsat."[82] The media also reported that the compromise included the removal of a provision contained in the House-passed bill, which would have allowed Level 4 "direct access" in Intelsat by U.S. companies before Intelsat became fully privatized.[83] In addition, the compromise legislation reportedly set an October 1, 2000, deadline for an IPO by Intelsat. The legislation would reportedly allow the FCC to extend that date to December 31, 2001, if privatization has not progressed sufficiently.

European and Administration Concerns

A preliminary version of the conference report placed conditions on FCC approval of applications by Intelsat, Inmarsat, and their successor entities for authorization to provide

---

[82] "Bliley, Burns, Markey Reach Agreement on Satellite Privatization Bill," *PRW via Dow Jones,* February 17, 2000.

[83] Mark Wigfield, "House, Senate Conferees Reach Accord on Comsat Bill," *Dow Jones Newswires,* February 17, 2000.

"non-core" services (such as Internet, high-speed data transmission and video services) to, from, or within the United States. These conditions were designed to ensure that competition in U.S. telecommunications markets is not harmed by privatizing Intelsat and Inmarsat. Unless the FCC finds that privatization is consistent with specific criteria, the FCC is required to determine that competition in U.S. markets will be harmed.

While negotiations between House and Senate conferees were on-going, two members of the European Union (EU) Commission (hereafter, the "EU Commission") expressed concerns that the Act would unfairly penalize European telecommunications companies by requiring Intelsat, Inmarsat, and their privatized and separated entities to meet certain criteria before entering the U.S. market. These individuals questioned whether such market access conditions are compatible with U.S. international obligations. They argued that imposing conditions on market access that applied only to the privatized successors of Intelsat and Inmarsat, and their privatized or spun-off entities, would violate the U.S. World Trade Organization (WTO) Basic Telecommunications Agreement obligations on market access, Most Favored Nation (MFN) status, and national treatment.[84]

Prior to signing the bill into law, the President Clinton's Administration raised concerns similar to those raised by the two EU Commission members. Most notably, the Administration questioned whether the bill would interfere with presidential authority to conduct diplomatic and foreign affairs, and potentially violate agreements of the WTO and the International Telecommunications Union (ITU). The Administration also contended that the legislation could result in limiting consumer choices and reducing U.S. market competition.

The *ORBIT Act* Becomes Law

The conference report (H.Rept. 106-509) passed the Senate by unanimous consent on March 2, 2000, and the House by a voice vote on March 9, 2000. On March 17, 2000, the President signed the *ORBIT Act* into law (P.L. 106-180). The final version of the Act amended the 1962 *Communications Satellite Act,* with the stated purpose of encouraging the privatization of the intergovernmental satellite organizations, Intelsat and Inmarsat, and promoting a competitive satellite communications marketplace. To accomplish this goal, the Act includes provisions to ensure that the privatized successors to Intelsat and Inmarsat be independent of signatories to either organization. It also seeks to promote privatization by offering the incentive of access to the U.S. marketplace if both organizations privatize in an expeditious and "pro-competitive" manner. If Intelsat and Inmarsat do not privatize in such a manner by a certain date, the Act limits their access to the U.S. market. To allow time for a reasonable transition, it gives Intelsat until April 1, 2001, and Inmarsat until April 1, 2000, to accomplish full privatization. The Act also requires the FCC to limit, deny, or revoke authority for either organization to provide "non-core" services to the U.S. market if the FCC determines that the organization has not privatized by these dates. Finally, the Act adds a number of deregulatory measures designed to ensure that all U.S. satellite service providers

---

[84] The United States is subject to market access conditions and other obligations toward foreign telecommunications service providers under the Fourth Protocol to the General Agreement on Trade in Services (GATS), which is generally referred to as the WTO Basic Telecommunications Agreement.

compete as efficiently as possible within the U.S. satellite marketplace. Most notably, it prohibits the FCC from auctioning orbital slots or spectrum assignments for global satellite systems, and requires the Administration to oppose such auctions in all international fora.

In signing the Act, President Clinton addressed some of his and the EU Commission's concerns by stating that the Administration intends to pursue Intelsat's and Inmarsat's privatization in a manner compatible with U.S. international obligations.[85] The President stated his plans to interpret provisions of the Act directing the Administration to take particular positions in international organizations as "advisory," and as not interfering with his "constitutional authority over foreign affairs." Moreover, President Clinton stated that he plans to continue to advise the FCC on the interpretation of and compliance with U.S. WTO commitments. The President also asserted that he will interpret the Act to require the FCC to make only one determination *in* issuing a license or other authority to a separated entity, and that it is his understanding that the Act does not limit the FCC's authority to auction spectrum and orbital slots.

It appears that the EU Commission's concerns regarding the Act's possible violation of the WTO Basic Telecommunications Agreement may have been met by both the inclusion of an additional clause in the final law, and by the President's statements regarding the Administration's interpretation of the law. The additional clause, Section 601(c), directs the FCC to make its determinations and licensing decisions "in a manner consistent with United States obligations and commitments for satellite services" under the WTO Basic Telecommunications Agreement.

MOST RECENT DEVELOPMENTS

Since the *ORBIT Act* became law, the FCC and other relevant federal agencies have begun to implement some of the Act's provisions. At the same time, some of the corporations and organizations most affected by the Act have also begun or continued to incorporate changes in their structures and policies, some directly related to the Act's provisions. The most significant and relevant of these events are summarized below.

- Since its September 1999 ruling, the FCC has granted permission to "directly access" Intelsat—without going through Comsat as an intermediary—to at least three U.S. telecommunications providers: Hughes Global Services, Inc., Spacelink International L.L.C., and British Telecommunications Broadcast Services North America.[86] Some analysts predict that such decisions will allow U.S. providers of international communications links to more cheaply and easily use Intelsat's satellites.

- On March 1, 2000, Comsat boosted its ownership in Intelsat by 2.1%, to a total of 22.5%, which will remain intact until February 28, 2001. According to news reports, Intelsat ownership is adjusted each year to reflect usage of the service by each of the

[85] The White House, Office of the Press Secretary, "Statement by the President on Reorganization of Intelsat," March 17, 2000.
[86] "Hughes has received FCC permission to provide customers with 'direct access,'" *Communications Daily,* April 13, 2000, p. 14; "U.S. End to Comsat Monopoly Eases Intelsat Satellite Access," *Space News,* June 12, 2000, p. 22.

143 signatories.[87] Comsat will reportedly pay an undetermined amount to other signatories to account for the value of the additional ownership interests. The true value of Intelsat has not yet been determined, since the organization will not be privatized until 2001 and cannot be taken public until after that event.

- Lockheed Martin announced that it would consider selling its satellite assets after purchasing the remaining shares of Comsat. A company spokesperson stated that the company is "more interested in developing a global telecommunications provider" than in "developing into a major satellite operator."[88] Therefore, the company is considering whether to retain Lockheed's existing satellite assets or to sell them to fund other efforts.

- News reports stated that Inmarsat has still not met all the conditions "demanded by U.S. authorities" in order to do business in the United States.[89] Among these not-yet-met conditions are a diluted ownership structure and the appointment of a board of directors that is independent of the treaty organization's former signatories. Inmarsat officials have reportedly stated that both of these conditions will be met when the company proceeds with an IPO of stock in mid-2001. However, some analysts maintain that provisions in the *ORBIT Act* may prevent the FCC from approving the sale of Inmarsat's services in the United States prior to meeting these conditions.

- Following requirements of the *International Anti-Bribery and Fair Competition Act of 1998* (P.L. 105-366), the Department of Commerce completed a report for the Organization for Economic Cooperation and Development (OECD), that included determining whether signatories to Intelsat and Inmarsat are accorded market advantages, such as immunities and market access in nations or regions served by either intergovernmental satellite organization. To prepare for the report, the Department sought comments from industry on the progress of Intelsat and Inmarsat signatories in providing full and open competition in satellite markets.[90] The report concluded that although international satellite organizations have, in the past, enjoyed advantages through the use of privileges and immunities, such advantages appear to be diminishing due to privatization, increased competition, and a global trend toward open markets. Specifically, the report stated that the *ORBIT Act* has provided a "vehicle to monitor the extent to which privatization reduces the advantages traditionally accorded international satellite organizations."[91]

[87] "Comsat Hikes Ownership in Intelsat," *Satellite Today,* April 3, 2000, p. 6.
[88] Sam Silverstein, "Lockheed to Consider Selling Satellite Assets After Merger," *Space News,* April 3, 2000; "Lockheed Martin Global Telecommunications Unit Plans Shift Toward Terrestrial Service," *Satellite News,* May 8, 2000, p. 1.
[89] Peter D. deSelding, "Private Inmarsat Stumbles on U.S. Market Hurdles," *Space News,* April 17, 2000, p. 3.
[90] "Dept. of Commerce is seeking comment," *Communications Daily,* April 20, 2000, p. 9; Department of Commerce, "Market for Satellite Communications and the Role of Intergovernmental Satellite Organizations," *Federal Register,* vol. 65, no. 75, April 18,2000, pp. 20804-20805.
[91] Department of Commerce, Trade Compliance Center, "Addressing the Challenges of International Bribery and Fair Competition (Second Annual Report to Congress on the OECD Antibribery Convention)," July 2000, p. 99, [http://www.mac.doc.gov/FCC/BRIBERY/oeed_report_2000/index.html], visited October 4, 2000.

- The Department of State sought similar comments, via the Department of Commerce notice, to meet specific *ORBIT Act* requirements. Under provisions in the *ORBIT Act*, the President is to provide an annual report to Congress on the progress of privatization in relation to the objectives, purposes, and provisions of the Act. Comments were due by May 8, 2000, but the report has not yet been issued.

- Also related to *ORBIT Act* requirements, the FCC issued a Notice of Proposed Rulemaking in June 2000.[92] The FCC stated that Intelsat must "redefine existing relationships with signatories and direct access users" as part of its move to privatization.[93] Furthermore, the FCC stated that it would be "concerned" if Comsat's control of space-based market services "effectively denies users the benefits of direct access," or if Comsat tries to increase its control of the space-based service market in order to "deny its availability" to other competitors. For these reasons, the FCC followed *ORBIT Act* guidelines and requested information and comments on whether users or service providers of telecommunications services have sufficient opportunities to access Intelsat's space-based services to meet their own service and capacity requirements. The notice also sought Comments regarding potential action by the FCC should it conclude that sufficient opportunity does not exist for users and service providers to "directly access" Intelsat. Comments were due on June 23, 2000.

- The FCC issued their findings on September 19, 2000, concluding that U.S. users and providers of telecommunication services currently do not have sufficient opportunity to access Intelsat's services to meet their own service or capacity requirements. Furthermore, the FCC stated that both Comsat and direct access users have reported difficulty in obtaining Intelsat's services to satisfy customer needs. These difficulties are reportedly primarily due to capacity shortages caused by high demand, and by "what appears to be procedural complications in 'matching' capacity that is available."[94] However, the FCC maintained that commercial negotiations are likely to be more successful in resolving these difficulties than a regulatory approach. Therefore, the FCC directed Comsat to continue work with U.S. telecommunications providers, and required Comsat and directly access users to the reports with the FCC by March 13, 2001, on the progress of these negotiations. The FCC stated that it may then "revisit this [commercial negotiation] conclusion depending on the outcome of the negotiation period."[95]

- In a subsequent report to Congress, the FCC stated that as the deadline for finalization of the privatized company draws near, Intelsat has yet to make major decisions on its operations.[96] Specifically-cited decisions that need to be made include: Intelsat's corporate structure and rules of operation; the date of an IPO; the

---

[92] Federal Communications Commission, "Availability of Intelsat Space Segment Capacity to Direct Access Users," *Federal Register*, vol. 65, no. 107, June 2, 2000, pp. 35312-35314.
[93] "FCC Focuses on Intelsat Privatization," *Communications Daily*, May 25, 2000, p. 5.
[94] Federal Communication Commission, "Availability of Intelsat Space Segment Capacity to Users and Service Providers Seeking to Access Intelsat Directly," Report and Order, FCC #00-340, September 19, 2000, p. 16.
[95] *Ibid.* p. 18.
[96] "Intelsat Slow to Determine Corporate Structure, FCC Says," *Communications Daily*, June 21, 2000, p. 3.

headquarters location; the location for taxes and satellite registration; connectivity arrangements with "lifeline users" (e.g. underdeveloped countries); and post-privatization distribution agreements and the role of Intelsat or an international treaty organization in supplying these lifeline services.

- Intelsat's private spin-off, New Skies, asked the FCC for a 6-month extension to the *ORBIT Act's* July 1, 2000 deadline to hold an IPO[97] Reportedly citing "market conditions and related business factors," the company stated that it would move forward with an IPO once market conditions improve. Though both Lockheed Martin and Inmarsat supported the request, PanAmSat voiced its objections. In an FCC filing opposing the bid for an extension, PanAmSat reportedly stated that a "pattern has emerged" in which New Skies has taken advantage of special privileges it has as an Intelsat signatory by "shielding information related to the proposed IPO" from the public and FCC review.[98] A PanAmSat spokesperson argued in reply comments to the FCC that the New Skies IPO should not be delayed beyond September 30, 2000.[99] Despite these objections, in July 2000 the FCC granted New Skies' request for an extension. Under new regulations, New Skies has until January 1, 2001 to conduct an IPO. The FCC reportedly stated that there was "no evidence" that New Skies had ulterior motivations in delaying the IPO, and rejected PanAmSat's recommendation to limit the extension to three months, saying that "best timing can't be accurately forecast."[100]

- In early August 2000, the FCC approved Lockheed Martin's acquisition of the remaining shares of Comsat, citing a "lack of potential competitive harm" and "beneficial efficiencies."[101] A few days later, Lockheed Martin acquired the remaining 51% stake of Comsat through a one-for-one, tax-free exchange of stock valued at $790 million.[102] Following *ORBIT Act* provisions, Comsat will continue to serve as the U.S. signatory to Intelsat and Inmarsat until both organizations are fully privatized, at which point Comsat will become a shareholder in both corporations.

- Intelsat, after recanting earlier statements that it might show its opposition to the Act by leaving the United States, applied to the FCC to become a U.S. licensee following privatization. In a separate FCC filing, PanAmSat objected to a provision of Intelsat's request regarding the transfer of five unoccupied (e.g. "warehoused") orbital slots to its private successor.[103] In August 2000, the FCC approved Intelsat's request for licenses for 17 existing and 10 planned satellites. Many analysts see the FCC's decision to give Intelsat the authority to operate in the United States as an important step in Intelsat's planned transition from an international treaty organization to a private corporation. Although three of the four members of the FCC

---

[97] "New Skies asked FCC for 6-month extension," *Communications Daily,* June 2, 2000, p.11.

[98] "PanAmSat said New Skies request to FCC for 6-month extension," *Communications Daily,* June 9, 2000, p. 10.

[99] "Satellite Companies Look Forward to Direct Access," *Communications Daily,* June 12, 2000, pp. 3-4.

[100] "New Skies Receives 6-month FCC Extension for IPO," *Communications Daily,* July 3, 2000, p. 7.

[101] "FCC Approves Comsat-Lockheed Martin Merger," *Communications Daily,* August 1, 2000, p. 3.

[102] "Lockheed completed Comsat buy, eyes more deals," *Aerospace Daily,* August 4, 2000, p. 1; "Lockheed Martin Finalizes Purchase of Comsat, Ending Its 38-year Existence," *Satellite News,* August 7, 2000, p. 2.

[103] "Intelsat shouldn't be allowed to transfer warehoused," *Communications* Daily, July 29, 2000, p. 11.

voted in favor of the license, Commissioner Harold Furchtgott-Roth voted in partial dissent. Specifically, Commissioner Furchtgott-Roth opposed a provision of the ruling that granted Intelsat waivers on technical specifications that other satellite operators must follow.[104] Opposing the decision, PanAmSat filed a petition with the FCC for reconsideration of the order.[105] A PanAmSat spokesperson reportedly stated that the FCC decision violated the "spirit" of the *ORBIT Act,* and that the technical waivers would give Intelsat an unfair competitive advantage.[106]

- Intelsat officials stated that the company plans to complete its privatization plan by April 2001.[107] The organization's signatories are to meet in November 2000 to give final approval to this measure.[108]

- Several U.S. telecommunications providers, including Comsat, filed applications with the FCC to utilize Inmarsat's "non-core" services once it becomes fully privatized. Opposing these requests, PanAmSat asked the FCC to deny Comsat the authority to use Inmarsat's space-based capabilities to provide U.S. domestic land mobile services.[109] Though the public comment period has closed, the FCC has not yet ruled on these applications. The FCC has indicated that it does not intend to rule on these applications until after Inmarsat meets specific privatization criteria outlined in the *ORBIT Act,* including changing the composition of its board of directors.

- Inmarsat—which recently changed its legal name from Inmarsat Holdings to Inmarsat Ventures[110]—also asked the FCC to extend its IPO deadline beyond the *ORBIT Act's* October 1, 2000 date. On September 29, 2000, the FCC granted the request for a nine-month extension.

- Lockheed Martin agreed to sell one-third of its interests in Inmarsat Ventures to Norway's Telenor."[111] Analysts expect the sale to generate more than $100 million for Lockheed Martin Global Telecommunications this year, and to reduce Lockheed Martin's equity in Inmarsat from slightly more than 22% to about 14%. The sale is expected to increase Telenor's share in Inmarsat from 6.8% to slightly more than 15%.

---

[104] For example, the FCC normally only grants licenses for satellites separated by a 2-degree orbital arc from other spacecraft. Some Intelsat satellites are closer or further away than 2 degrees. Commissioner Furchtgott-Roth's legal advisor, Bryan Tramont, stated that exceptions to this and other rules should be granted throughout the industry, and not only to Intelsat. For more information, see: "FCC Approves 27 Intelsat Licenses," *Space News,* August 14, 2000, p. 4.

[105] "PanAmSat Asks FCC to Reconsider Ruling to Allow Intelsat to Privatize as a U.S. Company," *Satellite News,* September 11, 2000, p. 2.

[106] "PanAmSat remains major opponent of FCC order," *Communications Daily,* August 16. 2000, p. 7.

[107] "Intelsat Chooses U.S. for Privatized Company Locale," *Satellite News,* September 18, 2000, p. 6.

[108] Michael A. Taverna, "Satellite Operators Skeptical About IGO Privatizations," *Aviation Week & Space Technology,* September 18, 2000, p. 32.

[109] "Inmarsat wants FCC to extend Oct. 1 deadline," *Communications Daily,* August 23, 2000, p. 9.

[110] "In effort to reinforce its strategy," *Communications Daily,* July 19, 2000, p. 12.

[111] "Lockheed Martin to reap over $100 million for sale of Inmarsat *shares,"* *Aerospace Daily.* September 14, 2000, p. 398; "Lockheed Sells 1/3rd of Inmarsat Equity," *Satellite News,* September 18, 2000, p. 7.

- The EU Commission continued to monitor both Intelsat's and Inmarsat's FCC application processes. Some analysts have predicted that the EU Commission might challenge the *ORBIT Act* through the WTO if the FCC declined the applications, or conditioned their approval, in terms unfavorable to the EU Commission.

APPENDIX A: THE CREATION AND EARLY REGULATION OF COMSAT, INTELSAT, AND INMARSAT

Material for this historical summary of Comsat's, Intelsat's, and Inmarsat's origins and early regulatory changes draws largely from three primary sources:

- Joseph N. Pelton, "The History of Satellite Communications," Chapter 1 in *Exploring the Unknown—Selected Documents in the History of the U.S. Civil Space Program, Volume II: Using Space,* John M. Logsdon et al., ed. (Washington: GPO: 1998), pp. 1-11.

- U.S. Congress, House Committee on Science and Technology, Subcommittee on Space Science and Applications, "Communications Satellites," Chapter 3 in *United States Civilian Space Programs, Volume* II: *Applications Satellites,* committee print prepared by the Congressional Research Service, Library of Congress, 98th Congress, 1st Session (Washington: GPO, 1983), pp. 11-52, 78-92, 120-144.

- U.S. Congress, Office of Technology Assessment, "Satellite Communications," Chapter 6 in *International Cooperation and Competition in Civilian Space Activities* (Washington: GPO, 1985), pp. 145-249.

Early Initiatives and Congressional Action

The launch of Sputnik 1 by the Soviets in October 1957 triggered several United States space initiatives, including those in satellite communications. In the late 1950s and early 1960s, communications satellite research and development programs were established within the National Aeronautics and Space Administration (NASA) and the Department of Defense (DoD). These programs assisted in establishing U.S. satellite communications capability, and were paralleled by private sector activities. At the same time, outmoded telegraph cables were being replaced by new submarine telephone cables across the Atlantic and Pacific Oceans, stimulating the growth of international telecommunications. Recognizing that high-capacity satellites could lead to rapid growth of global communications, NASA, DoD and American Telegraph and Telephone Company (AT&T) moved ahead with communication satellite projects, achieving several successful launches and advancing the commercial viability of satellite communications. A global network of stabilized satellites in geosynchronous orbits evolved shortly thereafter.

Policy decisions regarding the most appropriate method to institutionalize the civilian communications satellite system did not arrive as easily as the technology. During 1961 and

1962, there was intense debate in the United States regarding public versus private ownership and operations. The issue was featured prominently in the public when, in his May 25, 1961 speech to Congress, President Kennedy called for almost $10 billion over six years to support a robust space program. The program he called for included an "accelerated use of space satellites" for worldwide communications. The new President also called for a "constructive role for the U.N.[112] In international space communications.

Congress began discussing how to respond to Kennedy's proposal for a national space program in the latter half of 1961. Key among these discussions was the role that AT&T might play in a future communications satellite industry. Although AT&T enjoyed strong support from the FCC, several Members of Congress were wary of its power and monopolistic standing in telecommunications. Hearings were held regarding potential AT&T monopolization of any satellite system left in the hands of the telecommunications carriers. While there was agreement in Congress on the purposes and scope of a communications satellite system, controversy still surrounded its ownership and operation. Some argued that existing private companies would be the most effective developers and operators, as the United States already had the most efficient and least expensive communications system in the world. Others agreed that private enterprise was probably the most efficient path, but felt that the purposes of the system and the dangers of monopolization by existing private firms— most notably AT&T—dictated the need for a new approach: a public corporation. Still others felt that because the development of the technology was already at public expense, and because any satellite system should demonstrate U.S. capabilities and good will, the Government should own and operate the system. These concerns over the potential monopolization of satellite services by AT&T in the early 1960s were echoed in the compromise legislation, which was ultimately developed.

Three bills on this issue, paralleling the positions above, ultimately appeared in Congress. The first, introduced by Senator Kerr, sought to amend the *NASA Act of 1958* by creating a satellite communications corporation owned and operated exclusively by U.S. telecommunications companies providing international services.[113] The second bill, intended to codify the current Administration policy, was introduced by Senators Kerr and Magnuson in the Senate, and by Congressman George Miller in the House. This legislation called for the creation of a profit-making quasi-public corporation to operate the communications satellite system, with controlling voting stock held by the public. Then, in the midst of conflicts between Kerr and the Kennedy Administration over the role of the carriers in ownership and operation of the satellite system, Senator Kefauver introduced legislation providing for Government ownership of the system. Senator Kefauver argued that taxpayers had financed the technology and should, therefore, retain ownership. He also asserted that a private company would inevitably be dominated by AT&T, and that the process of creating a private company would slow development of an operational system. After some controversy and parliamentary difficulties, a bill resembling the then current Administration policy was sent to the President on August 27, 1961.

[112] *Public Papers of the Presidents of the United States, John F. Kennedy, 1961* (Washington: GPO. 1962), pp. 529-31, quoted in: Joseph N. Pelton, "The History of Satellite Communications," p. 4.
[113] Public Law 85-568; 42 U.S.C. § 2451 *et seq.*

The *Satellite Act* Creates Comsat

The Comsat Corporation was created on August 31, 1962, when President Kennedy signed the *Satellite Act*. The Act defined Comsat as a private corporation, which would serve as the only U.S. provider of satellite telecommunications. Comsat was to operate under a broad international charter, instead of one which served only high profit areas, so as to provide for a system which would, by definition, be accessible to all nations—including those to which service would show no immediate profit. The corporation was organized so that no single company could dominate the system through ownership or patent control, and the public would own 50% of the corporation's stock, with international carriers owning the remaining 50%. The Act emphasized the international mission of Comsat, and reserved domestic satellite operations for future consideration.

Comsat was incorporated in the District of Columbia in 1963. The *Satellite Act* provided that Comsat would exclusively represent the U.S. in the international satellite communications marketplace, though the Act was unclear regarding who would operate the U.S. domestic satellite system. In addition, while Comsat was a private profit-seeking entity, the Act only allowed the corporation to seek profits in the commercial satellite communications industry, and left unclear whether it could diversify into other lines of activity. The Act also provided for the corporation to ultimately represent U.S. interests within whatever international structure was established to manage the global system.

An International Treaty Creates Intelsat

At the time of passage of the *Satellite Act,* one objective of the United States, as stated in the Act, was to contribute to world peace and understanding through the development of a worldwide satellite communications system established in conjunction and cooperation with other countries. Therefore, shortly after its formation, Comsat officials and appropriate U.S. government agencies, acting under a congressional mandate to establish an international satellite communications system, began negotiations aimed at establishing an international organization to develop and manage a global commercial communications network. During these negotiations, U.S. representatives emphasized that such a satellite system should be commercially viable and technologically efficient. An international interim agreement signed by President Kennedy in 1964 created the emerging global system known as the International Telecommunications Satellite Consortium (lntelsat), with Comsat as its manager. After starting operations with the Early Bird satellite in 1965, Intelsat launched spacecraft to cover the three major ocean regions and established the first global satellite communications system in 1969. This civilian system began with an initial satellite stationed over the Atlantic Ocean in1965, then another over the Pacific Ocean in 1967, and, finally, a third over the Indian Ocean in 1969.

Intelsat was created to establish and maintain a global commercial telecommunications satellite system, and represented the first attempt to exploit commercial space technology on an international basis. The consortium was to operate under the interim arrangements for five years, after which time definitive agreements were negotiated. Original agreements provided that ownership be apportioned according to projected use of the system. The U.S. originally had 61% of the votes, and Comsat was designated as Intelsat's manager responsible for the design, development, construction, establishment, operation, and maintenance of the space segment. However, the desire by other nations to play a greater role in Intelsat led members of the system to decrease the dominant position of the U.S. in the definitive agreements.

The final agreements, which became effective via international treaty in 1973, outlined Intelsat's legal standing as an international intergovernmental organization (IGO), clarified its authority to provide domestic as well as international telecommunications services, and created procedures to ensure that member countries not establish separate, competing satellite systems. Intelsat began as a cooperative organization between governments and their signatories: nominee organizations that were, inmost cases, the country's PTT. All entities contributed the capital and shared the risk involved. The 1973 agreements also diminished the power of the U.S. within the Intelsat organization, and made the present organization broader and more representative of the views of the smaller nations.

The Subsequent Creation of Inmarsat

The International Maritime Satellite Organization (Inmarsat) came into existence through international agreements in July 1979. Inmarsat grew out of an initiative of the then International Maritime Consultative Organization, now the International Maritime Organization (IMO). Inmarsat arose from the IMO's requirement to provide vessels at sea with emergency communications. Therefore, Inmarsat's original purpose was to provide communications for commercial, distress and safety applications for ships at sea. At the time, technology for mobile satellite communication was somewhat unexplored, and tile viability of the industry was unproven. Therefore, like Intelsat, Inmarsat began as a joint cooperative venture between governments and their signatories, which were, for the most part, the signatory country's PTT. Also like Intelsat, all entities contributed the capital and shared the risk involved. Comsat's functions for Inmarsat were codified by the 1978 amendments to the *Satellite Act,* which specified that Comsat would serve as the U.S. signatory to Inmarsat.[114] Similar to the Intelsat-Comsat situation, Comsat has financial interest in Inmarsat facilities.

Early Regulatory Changes

Since the authority to regulate was vested in the FCC by the *Satellite Act,* regulatory changes in the satellite communication market have occurred, in large part due to responses to growth in the industry. For example, beginning in 1966, several companies applied to the FCC for licenses to operate domestic telecommunications satellites. A July 1972 decision by the FCC allowed the use of satellites for domestic communications within the United States, and thus, opened a large new market for satellite telecommunications.[115]

---

[114] Public Law 95-564; 47 U.S.C. §751 *et seq.*

[115] Federal Communications Commission, Federal Communications Commission Reports: Decisions and Reports of the Federal Communications Commission of the United States, June 9, 1972 to August 4, 1972, Volume 35, Second Series (Washington: GPO, 1974), pp. 844-85 1, 860-867, in: *Exploring the Unknown: Selected Documents iii the History of the U.S. Civil Space Program, Volume II: Using Space,* pp. 120-132.

Then, in the mid 1980s, the FCC advocated separate and competing international satellite systems. In doing so, the FCC encountered considerable hostility from the international satellite organization signatories, most of whom stated that, if the FCC permitted satellite systems separate from Intelsat and Inmarsat, that they would not cooperate with these systems. Despite these statements, several companies applied to the FCC for licenses to provide international telecommunications satellites. Applicants included Orion Satellite Corporation, International Satellite, Inc., Cygnus Satellite Corporation, RCA Communications, Inc., and Pan American Satellite (PanAmSat) Separate Systems. Intelsat disagreed with these applications, stating that the negotiators of the Intelsat agreements never contemplated the entrance of separate systems and that such systems would cause significant economic harm."[116] The organization asserted that it should be the sole arbiter of whether the separate systems should be allowed. The FCC ultimately rejected Intelsat's position and granted licenses for separate systems, subject to restrictions.[117] This policy decision is known as the "Separate Satellite System Policy," or "SSP." Initially, the separate systems were prohibited from carrying basic voice traffic, preserving that market exclusively for Intelsat and Inmarsat. This limitation on voice traffic, however, has since been removed. However, the separate systems have continued to focus primarily on the development of the video market.

[116] Chris Rourk. "Analysis of the Technical and Economic Issues Raised in the Consideration of International Telecommunications Satellite Systems Separate from Intelsat," *Federal Communications Law Journal,* March 1994, Vol.46, pp. 329-346, information from p. 332.
[117] The restrictions limited the separate systems to provision of services through the sale or long-term lease of transponders. Until 1997, the separate systems could not connect with the public-switched network.

Chapter 7

TELEMARKETING: DEALING WITH UNWANTED TELEMARKETING CALLS

James R. Riehl

Telephone marketing, better known as telemarketing, was an approximately $669 billion business in the United States, according to the Direct Marketing Association (DMA). As telephone technologies become mote advanced, it becomes more and more cost-effective for telemarketers to use these technologies to sell various products and services directly to consumers. There are only a few households that have not received a telemarketing call. However, with the expansion of the telemarketing business comes the expansion of telemarketing fraud. Although the vast majority of telemarketers are legitimate business people, there are other individuals and companies who violate existing laws and rules and bilk unsuspecting customers of $40 billion a year according to some estimates.

Table 1. Telemarketing in the United States
(dollars in billions)

	1995	1999	2000	2001	2005
Consumer	$167.0	$236.0	$256.9	$276.6	$373.3
Business-to-Business	$200.2	$317.60	$354.7	$392.2	$566.3
Total	$367.2	$553.6	$611.7	$668.8	$939.5

Source: Direct Marketing Association. *2000 Economic Impact: U.S. Direct Marketing Today.* [http://www.the-dma.org/library/publications/charts/dmsales_medium_market.shtml]

FEDERAL LAWS

In recent years, seven federal laws that deal directly with telemarketing issues have been enacted. Each of the laws has a different focus relative to telemarketing, but all attempt to limit or prohibit certain abusive, fraudulent, or deceptive practices. Because both the Federal Communications Commission (FCC) and the Federal Trade Commission (FTC) are responsible for different aspects of telemarketing, both agencies were directed to promulgate regulations. The FCC generally covers consumers' privacy rights relative to telemarketing practices and the use of the telephone system by telemarketers to transmit information whether via facsimile machine, automated dialing mechanisms, recorded calls, or by a live person. The FTC is concerned more with the content and consequences of the call, and its regulations focus on whether certain sales practices are misleading, fraudulent, or deceptive. Individual states have passed laws relating to telemarketing practices within a state.

Telephone Consumer Protection Act of 1991 (TCPA)

The TCPA, P.L. 102-243, was signed by President Bush on December 20, 1991. It was the first of the six federal laws passed dealing specifically with telemarketing issues. This Act directed the Federal Communications Commission to issue rules balancing the fair business practices of telemarketers with the privacy concerns of consumers.

Some of the provisions of the FCC rules resulting from this Act are:

(1) Telemarketing companies must maintain a do-not-call list for calls placed to residential telephone numbers. If a consumer requests that his/her name be placed on such a list, the company must honor the request for 10 years. Nonprofit and charitable organizations are exempted from this provision, and the rules do not apply to calls placed to business telephone numbers.

If a consumer's name is on a company's do-not-call list and the company places more than one call to that consumer in the year after the consumer has been placed on the list, the consumer may, if he/she wishes, sue the telemarketer in state court, usually a small claims court. Should a consumer pursue court action, he/she should maintain records of all calls and contacts with the company.

(2) Telephone solicitations to private residences may only be made between the hours of 8 a.m. and 9 p.m.

(3) Use of autodialers or prerecorded (artificial) voice messages to call any emergency telephone line (911, hospital, medical office, health care facility, poison control center, police, or fire lines), a guest or patient room in a hospital, health care facility, or home for the elderly, any phone number assigned to a paging service or cellular telephone, or services for which the person called would be charged for the call are prohibited unless prior consent was given to receive such calls.

(4) Prerecorded (artificial) voice calls to private homes are prohibited. However, such calls are permitted if the person called has consented to receive such calls, the call is noncommercial (from a charitable, nonprofit, political, or polling organization or government agency), the entity calling has an established business relationship with person called, or the call is an emergency. Such calls to business numbers are permitted.

(5) Any person or entity making a telephone solicitation to a private home must provide the name of the individual caller, the name of the person or entity on whose behalf the call is being made, and a telephone number or address where that person or entity may be contacted.

(6) Any person or business using autodialers or prerecorded (artificial) voice calls, including calls placed to businesses, must state its identity at the beginning of the message and its telephone number or address during or after the message.

(7) Autodialer calls may not lock onto a phone line. Within 5 seconds of a phone being hung up, the autodialer must release the phone line. In some areas of the country, due to different telephone system technologies, this release may take longer. Customers should check with their local telephone company for additional information.

(8) Calls transmitting unsolicited advertisements to home or business fax machines are prohibited unless permission has been granted to do so or there is an established business relationship. Any message sent to a fax machine must include the date and time the transmission is sent, identity of the sender, and the telephone number of the sender or the sending fax machine.

The FCC's initial final rule can be found in the *Federal Register* of October 23, 1992, on pages 483 33-36. After some modification, the rule was finalized in August 1995 and published in the *Federal Register* of August 15, 1995, on pages 42068-69. The Consumer information Bureau of the FCC provides more detailed information concerning the Telephone Consumer Protection Act of 1991 at its Web site: [http://www.fcc.gov/cib/consumerfacts/Nofaxes.html].

Telemarketing and Consumer Fraud and Abuse Prevention Act

P.L. 103-297 was signed by President Clinton on August 16, 1994. The Act directed the Federal Trade Commission to establish rules to prohibit certain telemarketing activities. The FTC's final Telemarketing Sales Rule (TSR) was adopted on August 15, 1995. The rule covers most types of telemarketing calls and also applies to calls consumers make in response to material received in the mail, but it is not intended to affect any state or local telemarketing laws. The rule went into effect on December 31, 1995.

Some of the provisions of the rule are:

(1) The rule restricts calls to the hours between 8 a.m. and 9 p.m.

(2) It forbids telemarketers from calling consumers if they have been asked not to. Violations of this provision may be reported to the state Attorney General.

(3) It requires certain prompt disclosures, prohibits certain misrepresentations and lying to get consumers to pay, and makes it illegal for a telemarketer to withdraw money directly from a checking account without the account holder's specific, verifiable authorization.

Telemarketers calling consumers must promptly identify the seller of the product or service that the purpose of the call is to sell something, the nature of the goods or services being offered and, in the case of a prize promotion, that no purchase or payment is required to participate or win. In addition, prior to a consumer paying for any good or service, the consumer must be provided with material information that is likely to affect their choice of the good or service. Material information includes cost and quantity; restrictions, limitations, or conditions; refund policy; and, in the case of a prize promotion, information on the odds of winning, that there is no payment required to enter the promotion, how the consumer may enter the promotion - without paying, and information on any material costs or conditions that may be required to receive or redeem any prize.

(4) Telemarketers and sellers are requited to maintain certain records for 2 years from the date that the record is produced. Records includes items such as advertising and promotional materials, information about prize recipients, sales records, employee records, and all verifiable authorizations for demand drafts for payment from a consumer's bank account.

There are exceptions to the Telemarketing Sales Rule. For example, calls initiated by a consumer that are not made in response to a solicitation, business-to-business calls (in most cases), and sales of 900-number (pay-per-call) services and franchises (covered by other FTC rules) are not covered. Certain types of businesses -- banks, federal credit unions, federal savings and loans, common carriers (long-distance telephone companies and airlines), non-profits, insurance companies, and many types of companies selling investments—are not covered by the rule because they are exempted from the FTC's jurisdiction and, in most cases, are regulated by other federal agencies. However, individuals or companies providing telemarketing services under contract for these companies are covered. The application of the rule can be complex. Consumers should check with state authorities or the FTC for clarification of coverage.

With certain limitations, the FTC, the states, and private individuals may bring civil actions in federal district courts to enforce the rule.

A statement of purpose and the text of the Telemarketing Sales Rule can be found in the August 23, 1995, *Federal Register,* pages 43842-77. The full text of the sales rule is also available at the Federal Trade Commission Web site at [http://www.ftc.gov/bcp/telemark/rule.htm]. Additional FTC information concerning telemarketing is available at [http://www.ftc.gov/bcp/menu-tmark.htm]. Complaints may be filed electronically at this site.

Telemarketing Sales Rule Review

During 2000, as required by this Act, the FTC began conducting a review of the rule's effectiveness, its overall costs and benefits, and its regulatory and economic impact since its adoption. In addition, the FTC examined telemarketing and its impact on consumers over the past 2 decades.[1] Results of the review will be reported to Congress. See {http://www.ftc.gov/bcp/rulemaking/tsr/tsr-review.htm] for further information.

---

[1] *Federal Register*, February 28, 2000, p. 10428-34.

On January 22, 2002, the FTC announced substantial proposed changes to the TSR. Among the FTC's proposed changes is the creation of a national do-not-call list. If enacted, consumers would place their name on the list by calling a toll-free telephone number. It would then be against the law for a telemarketer to call those consumers. Because certain businesses are exempt from the TSR (see above), notably those that consumers have given permission to contact them, placing a name on the list would not stop all telemarketing calls. Details concerning the procedures and operation of the national list will be addressed during review of the FTC proposal. If implemented, the national list will not be operational for many months.[2]

In addition, the FTC proposed that telemarketers be prohibited from blocking caller ID systems, that telemarketers be prohibited from obtaining a consumer's credit card or other account number from anyone but the consumer or from improperly sharing that number with anyone else for use in telemarketing, and that use of predictive dialers resulting in "dead air" (i.e., no one on the line) violates the TSR.

More detail concerning the FTC proposal may be found at the FTC Web site at: [http://www.ftc.gov/opaI2002/01/donotcall.htm]. The full text of the FTC Notice of Proposed Rulemaking is available at this site. In addition, the Notice was published in the *Federal Register* on January 30, 2002, on pages 4491-4546. The FTC will accept comments on the proposed TSR rule changes until March 29, 2002.

Senior Citizens Against Marketing Scams Act of 1994

This Act was Title XXV of the Violent Crime Control and Law Enforcement Act of 1994, P.L. 103-322, and was signed by President Clinton on September 13, 1994. It included provisions that increased penalties for telemarketing fraud against people over 55 years old. Provisions of this law allow imprisonment up to an additional 5 years for certain telemarketing crimes or up to 10 additional years if 10 or more persons over the age of 55 were victimized or the targeted persons were over 55. Also, the Act requires that full restitution be paid to victims and directs the U.S. Attorney to enforce any restitution order.

Telecommunications Act of 1996

P.L. 104-104, signed by President Clinton on February 8,1996, was a substantial amendment to the 62-year old Communications Act of 1934. Section 701 of the Act closed a loophole that allowed information service providers and telemarketers to connect callers to "pay-per-call" services even though the callers had initially dialed a toll-free telephone number.

---

[2] For additional information on do-not-call lists, see CRS Report RS21122, *Regulation of the Telemarketing Industry: State and National Do Not Call Registries*, by Angie A. Wellborn.

Telemarketing Fraud Prevention Act of 1997

P.L. 105-184 was signed by President Clinton on June 23, 1998, and attempts to deter fraudulent telemarketers by raising the federal criminal penalties for telemarketing fraud and permitting the seizure of a criminal's money and property to make restitution to victims. Also, if persons over the age of 55 were targets of the fraudulent telemarketing activities, criminals may be sentenced to additional prison time.

Protecting Seniors from Fraud Act

President Clinton signed the Protecting Seniors from Fraud Act, P1. 106-534, on November 22, 2000. This Act finds that an estimated 56% of the names on calling lists of illicit telemarketers are individuals aged 50 or older and that, as a result, older Americans are often the target of telemarketing fraud.

Among other things, this Act:

(1) Authorizes $1 million for each of the fiscal years 2001 through 2005 to be appropriated to the Attorney General (Department of Justice) for senior fraud prevention program(s) and directs the Comptroller General to submit a report to Congress on the effectiveness of the program(s).

(2) Directs the Department of Health and Human Services to provide and disseminate within each state information that both educates and informs senior citizens about the dangers of fraud, including telemarketing fraud. This information may be distributed via public service announcements, printed matter, direct mailings, telephone outreach, or the Internet.

(3) Instructs the Attorney General to conduct a study of crimes against senior citizens. Among other issues, the report must address the nature and extent of telemarketing fraud against seniors.

(4) Directs the Attorney General, not later than 2 years after the date of enactment of this Act, to include statistics relating to crimes against seniors in each National Crime Victimization Survey.

(5) Expresses the sense of the Congress that state and local governments should incorporate fraud avoidance information and programs into programs that provide assistance to the aging.

Crimes Against Charitable Americans Act of 2001

The Crimes Against Charitable Americans Act of 2001 was passed as Section 1011 of the Uniting and Strengthening American by Providing Appropriate Tools Required to Intercept and Obstruct Terrorism Act of 2001 (USA PATRIOT ACT), P.L. 107-56, and was signed by President George W. Bush on October 26, 2001.

Passed following the September 11 attacks, Section 1011 amends the Telemarketing and Consumer Fraud and Abuse Prevention Act (see above) to expand the coverage of the FTC's Telemarketing Sales Rule to apply to calls made to solicit charitable contributions. The rule currently covers only calls made to sell goods and services. Charitable organizations are exempt from the rule. Although this law does not remove that exemption, it does permit the FTC to take action against a for-profit company that fraudulently, deceptively, or abusively solicits charitable contributions for charities.

In addition, the Act increases the penalty for impersonating a Red Cross member or agent.

What Consumers Can Do

Hang Up

Simply say, "No, thank you. I'm not interested." No one has to make any special excuse or listen to the presentation of the person on the line.

Be Informed

Consumers can go to a local library and ask for help in finding information on telemarketing and telemarketing scams. The library's Internet connection (if available) will provide access to the Web sites listed in this report. If no Internet connection is available, one may contact the FTC to request information about the Telemarketing Sales Rule.

Federal Trade Commission
Public Reference Branch
6th Street and Pennsylvania Avenue, N.W.
Washington, D.C. 20580
[http://www.ftc.gov]

Contacting the local Better Business Bureau (BBB) to obtain information about telemarketing and the various types of telemarketing scams is another option. The BBB also provides information at its Web site [http://www.bbb.org/library/batele.html].

Be Cautious

If a letter or postcard arrives or there is a message on a home answering machine stating that someone has won a free trip or a prize or a sweepstakes, be cautious. Consumers should check the area code for the number that must be called to claim a prize or respond to a telemarketing call before they make the call. Most Caribbean countries and Canada have area codes that are integrated into the U.S. telephone system and may be reached by direct dialing without using separate country codes. Simply making a call to certain area codes may incur substantial long-distance charges. Those charges will depend upon the area code called, long-distance carrier, length of call, a customer's long-distance calling plan (or lack thereof), and other factors. 800, 877, 888, 866, and 855 are the toll-free telephone prefixes used in the United States.

A list of area codes and the state, territory, or country served by a particular area code may be viewed at the Web site of the North American Numbering Plan Administration (NANPA), the administrator of the North American telephone numbering plan. The Web site is [http://www.nanpa.com/area codes/index.html]. Also, the front section of a local phone book usually contains maps and listings identifying area codes. If a particular code is not listed, customers should call the phone company to determine what area an unfamiliar area code serves. According to an FCC Fact Sheet, there arc approximately 317 area codes in service today. About 207 of them are within the United States. Area codes are constantly being added or revised.

If people respond to a prize announcement, they should not give out credit card, bank account, or Social Security numbers or send money to cover taxes, customs fees, etc., unless they completely understand all charges, procedures, and details concerning the offer. There are many different types of telemarketing scams involving many different types of products and services. For example, some of the known scams deal with stocks and other investments, automatic debit, charitable donations, easy credit, credit cards, credit repair, advanced fee loans, magazine subscriptions, international telephone calls, prizes, sweepstakes, work-at-home schemes, and travel. If there is doubt, people can request that written documentation explaining the prize, product, or service be forwarded to them. Any reputable telemarketer will send the information. Consumers should take their time and not be pressured into responding immediately.

Report Incidents to the Authorities

If a consumer believes that lie/she is a victim of a telemarketing scam or that a telemarketing concern is violating existing rules, they should report the incident(s). First, contact a local or state consumer affairs office or the state attorney general's office. The FTC's Telemarketing Sales Rule permits local authorities to prosecute telemarketing seam operators who operate across state lines, and individual states may have passed their own laws or established regulations concerning telemarketing.

Federal Trade Commission
Federal authorities may also be contacted. Victims of false or deceptive telephone solicitation sales practices may file a written complaint with the FTC by sending a description of their situation to:

Federal Trade Commission
Consumer Response Center
Drop H285
Washington, D.C. 20580
1-877-FTC-HELP (382-4357)(toll-free)
An electronic complaint form and information are available at:
[http://www.ftc.gov]
Click on File a Complaint Online at the bottom of the screen.

The information that is provided to the FTC may help the agency establish a pattern of violations that may require action. However, the FTC generally does not get involved in individual disputes with telemarketing companies.

Project Know Fraud

In November 1999, President Clinton announced a new mail campaign designed to fight telemarketing fraud. The campaign, known as Project Know Fraud, mailed postcards about telemarketing fraud to every household in the United States and made a Know Fraud video available at libraries throughout the country. The project is a joint effort of the AARP (formerly known as the American Association of Retired Persons), Council of Better Business Bureaus' Foundation, Department of Justice, Federal Bureau of Investigation, Federal Trade Commission, National Association of Attorneys General, Securities and Exchange Commission, and U.S. Postal Inspection Service. For information on telemarketing fraud or to report suspected fraudulent telemarketing activity:

Know Fraud
P.O. Box 45600
Washington, D.C. 20026-5600
1-877-987-3728 (toll-free)
[http://www.consumer.gov/knowfraud]

National Fraud Information Center

Suspected fraudulent telemarketing activities may also be reported to the National Fraud Information Center (NFIC), a private, nonprofit organization that assists consumers with telemarketing complaints. NFIC forwards all appropriate complaints to the FTC. One may obtain information on telemarketing or report suspicious incidents to NFIC via telephone, mail, or the Internet.

National Fraud Information Center
P.O. Box 65868
Washington, D.C. 20035
1-800-876-7060
[http://www.fraud.org]

Federal Communications Commission

One may also contact the Federal Communications Commission if they believe violations of the Telephone Consumer Protection Act have occurred. Send a letter describing the complaint in detail to:

Federal Communications Commission
Consumer Information Bureau
Complaints
445 12th Street, N.W.
Washington, D.C. 20554
[http://www.fcc.gov/eib/consumerfacts/Nofaxes.html]

Ask to Be Placed on a Do-Not-Call List

If someone wants to be placed on a firm or individual's do-not-call list, they should state clearly and firmly to the caller that their name is to be added to the list. The caller must take the name, add it to their list, and keep it there for 10 years. In addition, consumers should take down the name of the caller, the name of the firm or individual for whom they ate making the call, and the address and telephone number where they can be reached. Note the date and time and keep a record of any additional calls that are received (if any) from the same source. If additional calls from the same source continue, a consumer may consider filing a suit in small claims court.

State Do-Not-Call Lists

Several individual states have passed laws establishing do-not-call lists within the state, Consumers should contact a local consumer affairs office, Better Business Bureau, or an appropriate state office to find out the particulars of their state's do-not-call list or if such a state list exists. In some cases, there may be a (monthly or annual) charge to be added to the list. If charges are assessed, state law, not FCC or FTC regulations, will determine the charges.

The FTC provides a list of states with do-not-call lists and contact phone numbers for those states at
[http://www.ftc.gov/bcp/conline/pubs/alerts/dnca1rt.htm].

Direct Marketing Association

The Direct Marketing Association, a national trade association serving the direct marketing industry, maintains a national do-not-call list. If someone adds their name to the national list, it will take a few months for it to take effect, and the name will go **only** to the companies who subscribe to the DMA's list service. Telemarketing companies are not required to subscribe to the service, and getting a name on any do-not-call list does not remove it from all telemarketers lists. The list is updated four times per year, in January, April, July, and October. There is no charge to add a name to the list. Send name, address, and home telephone number (including area code) to:

Telephone Preference Service
Direct Marketing Association
P.O. Box 9014
Farmingdale, NY 11735-9014
[http://www.the-dma.org/consumers/consumerassistance.html]

In some instances, telemarketers use automated dialing mechanisms that call every number in a targeted geographic area or with a certain prefix. If someone is in one of those areas or has the designated telephone prefix, they will not escape the call even if their name is on a do-not-call list. Even unlisted numbers are called in these situations. The DMA has established operational guidelines for its members using automatic dialing equipment and software.

The DMA also offers a free Mail Preference Service (MPS) for those who wish to receive less advertising mail at home. As with the Telephone Preference Service, to register for the MPS, a postcard or letter-providing name, home address, and signature must be sent to the DMA.

Mail Preference Service
Direct Marketing Association
P.O. Box 9308
Farmingdale, NY 11735-9008
[http://www.the-dma.org/consumers/consumerassistance.html]

DMA Privacy Promise

On July 7, 1999, the DMA announced implementation of its *"DMA Privacy Promise to American Consumers.* "This effort requires all DMA members to adhere to a set of consumer privacy protection measures. These measures include:

(1) Disclosing to consumers when contact information about them may be shared with other marketers;

(2) Providing a means for consumers to opt-out of any information sharing arrangement;

(3) Honoring any individual consumer's request not to receive any further solicitations from the marketer; and

(4) Requiring member companies to use the DMA's Mail Preference and Telephone Preference Services to maintain updated marketing lists of consumers who have chosen to place their name on these lists.

DMA members were given reasonable time to comply, and through the use of secret shoppers, decoys, review of consumer complaints, and random staff contacts, the DMA seeks to assure that its members comply with these measures. If a member refuses to correct its procedures when asked to do so, the DMA Board may expel the company and make its actions public.

In addition, the DMA, in conjunction with the FTC and FCC, has developed a Web page providing advice to consumers who shop by phone. The site provides shopping tips, information on federal laws and regulations, and provides information on filing complaints

[http://www.the-dma.org/consumers/shoppingbyphone.html].
The Direct Marketing Association
1120 Avenue of the Americas
New York, NY 10036-6700
(212) 768-7277

Chapter 8

DIGITAL SURVEILLANCE: THE COMMUNICATIONS ASSISTANCE FOR LAW ENFORCEMENT ACT AND FBI INTERNET MONITORING

Richard M. Nunno

SUMMARY

The Communications Assistance for Law Enforcement Act (CALEA, P.L. 103-414, 47 USC 1001-1010), enacted October 25, 1994, is intended to preserve the ability of law enforcement officials to conduct electronic surveillance effectively and efficiently despite the deployment of new digital technologies and wireless services that have altered the character of electronic surveillance. CALEA requires telecommunications carriers to modify their equipment, facilities, and services, wherever reasonably achievable, to ensure that they are able to comply with authorized electronic surveillance actions.

The modifications, originally planned to be completed by 1998, have been delayed due to disagreements among the telecommunications industry, law enforcement agencies, (led by the Federal Bureau of Investigation (FBI)), and privacy rights groups, over equipment standards, and other technical issues. The amount of federal funds to be provided to the telecommunications carriers for implementation of CALEA, which carriers are eligible to receive those funds, and concerns over privacy rights of individuals, have also impeded implementation.

After receiving petitions from the industry and the FBI over the dispute, the Federal Communications Commission (FCC) in 1999 ruled in favor of most of the FBI's requests to require carriers to implement upgrades to their telecommunications networks (The FCC did extend the deadline for some of the upgrades requested by the FBI). This decision resulted in lawsuits being filed by industry and privacy rights groups. In August 2000, a federal appeals court upheld parts of the FCC's decision, but remanded most of it back to the FCC for reconsideration.

CALEA originally authorized $500 million to be distributed to telecommunications carriers for implementation. After enactment, industry and federal Cost estimates increased, reaching $2 billion to $5 billion by 1999. Funding for CALEA was postponed for several years. To date, a total of $299 million has been appropriated to the Department of Justice to reimburse telecommunications carriers implementing CALEA, but has not yet been released.

The ongoing CALEA debate has found a renewed interest in Congress in connection with revelations about the FBI's efforts to monitor Internet communications using a computer system called Carnivore. Privacy rights issues related to Internet monitoring are similar to those in the CALEA debate because the format of data that can be collected from Internet Communications is increasingly being used in telephone communications. Privacy rights groups believe that recent court rulings overturning portions of the FBI's CALEA requirements will set a precedent for future Internet monitoring rules.

Given that CALEA has still not been implemented six years after enactment, and law enforcement's digital surveillance actions are increasing, some question whether CALEA was ever necessary. Others argue that other technologies may still be used to circumvent the ability of law enforcement to perform wiretaps.

BACKGROUND

In the early 1990s the Federal Bureau of Investigation (FBI) asked Congress for legislation to assist law enforcement agencies to continue conducting electronic surveillance. The FBI argued that the deployment of digital technologies in public telephone systems was making it increasingly difficult for law enforcement agencies to conduct electronic surveillance of communications over public telephone networks. As a result of these arguments and concerns from the telecommunications industry,[1] as well as issues raised by groups advocating protection of privacy rights,[2] the Communications Assistance for Law Enforcement Act (CALEA) was enacted on October 25, 1994 (47 USC 1001-1021), in the final days of the 103rd Congress.

CALEA is intended to preserve the ability of law enforcement officials to conduct electronic surveillance effectively and efficiently, despite the deployment of new digital technologies and wireless services by the telecommunications industry. CALEA requires telecommunications carriers to modify their equipment, facilities, and services to ensure that they are able to comply with authorized electronic surveillance. These modifications were originally planned to be completed by October 25, 1998. To date, however, implementation of CALEA is significantly behind schedule because the telecommunications industry, law enforcement agencies, and privacy rights groups have not reached an agreement on technical issues. Issues also remain concerning the amount of federal funds to be provided to the telecommunications carriers for implementation of CALEA, and the privacy rights of individuals that may be affected under various implementation scenarios.

---

[1] In this report, the telecommunications industry includes common carrier telephone companies, mobile wireless telecommunications providers, telecommunications equipment manufacturers, and other entities that provide telecommunications services to the public.

[2] Privacy rights groups involved in the CALEA debate include the Electronic Privacy Information Center, the Electronic Frontier Foundation, advocacy groups which both support on-line privacy rights of individuals, the Center for Democracy and Technology, which also advocates electronic privacy (and is funded primarily by the telecommunications, computer, and media industries), and the American Civil Liberties Union (ACLU), which represents a broad array of civil rights based on the First and Fourth Amendments.

Some Technical Terms

As a result of the revolution in digital technology in telecommunications, the process of wiretapping and other electronic surveillance has become more complex, and legal ambiguities have been introduced. As a background to understanding the problems associated with CALEA implementation, the definitions of several terms are necessary. Electronic surveillance refers to either the interception of communications content (as in a conversation) also known as *wiretapping*, or the acquisition of call-identifying information (the number dialed). The latter activity is accomplished through the use of *pen register devices*, which capture call-identifying information for numbers of outgoing calls from the location of lawful interception, and *traps and traces*, which capture information for numbers received at the location of lawful interception, much like consumer caller ID systems. Under current federal law, law enforcement (i.e., police or the FBI) must obtain a court order before conducting any of these activities. However, a wiretap requires a higher "evidentiary burden" than a pen register or trap and trace, including showing that there is probable cause for believing that a person is committing one of a list of specific crimes.[3]

Under traditional analog technology, it was easy to separate the above categories of electronic surveillance. However, the advent of digital signal transmission technologies has made that distinction less clear. Information signals (voice or data) can be transmitted over telephone networks in one of two ways: *circuit-switched* and *packet-switched* modes.[4] In circuit-switched systems, a communications path is established between the parties and dedicated exclusively to one conversation for the duration of the call. In packet-switched systems, the information is broken down into smaller pieces called "packets" using a digital process. Each packet contains a small part of the message content along with call-identifying information called a "header" that indicates the origination and destination points of the information. Each packet is transmitted separately and is reassembled into the complete message at the destination point.

The packet-switched mode is the signal transmission technology used in all Internet communications. Packet switching is considered a more efficient use of a network than circuit switching because the same line can be used for multiple communications simultaneously. Although the circuit-switched mode was historically used in all voice telephone calls, the packet-switched mode is increasingly being used for voice and data transmissions over telephone networks.

CALEA'S MAIN PROVISIONS

CALEA requires telecommunications carriers to assist law enforcement in performing electronic surveillance on their digital networks pursuant to court order or other lawful authorization. The telecommunications industry, privacy rights groups, and law enforcement

---

[3] See CRS Report 98-326 A, *Taps, Bugs & Telephony: An Overview of Federal Statutes Governing Wiretapping and Electronic Eavesdropping,* March 23, 1998.

[4] Switches are network devices that select a path or circuit for sending data to its next destination over the telephone network. Switches may also include functions of the router, a device also used in computer networks, that determines the route and adjacent network point for data to be sent.

agencies agree that CALEA was not intended to expand law enforcement's authority to conduct electronic surveillance. On the contrary, CALEA was intended only to ensure that after law enforcement obtains the appropriate legal authority, carriers will have the necessary capabilities and sufficient capacity to assist law enforcement in conducting digital electronic surveillance regardless of the specific telecommunications systems or services deployed.

CALEA (47 USC 1002) directs the telecommunications industry to design, develop, and deploy solutions that meet certain assistance capability requirements for telecommunications carriers to support law enforcement in the conduct of lawfully authorized electronic surveillance. Pursuant to a court order or other lawful authorization, carriers must be able, within certain limitations, to: (1) expeditiously isolate all wire and electronic communications of a target transmitted by the carrier within its service area; (2) expeditiously isolate call-identifying information that is reasonably available on a target; (3) provide intercepted communications and call-identifying information to law enforcement; and (4) carry out intercepts unobtrusively, so targets are not made aware of the electronic surveillance, and in a manner that does not compromise the privacy and security of other communications.

To allow carriers to give law enforcement the means to conduct its wiretaps, CALEA (47 USC 1003) requires the Attorney General to determine the number of simultaneous interceptions (law enforcement agencies' estimate of their *maximum capacity* requirements) that telecommunications carriers must be able to support. This action was originally required within one year of enactment, but was later delayed (see Initial Delays below).

To maintain privacy rights of individuals, CALEA (47 USC 1004) requires telecommunications carriers to ensure that any interception of communications or access to call-identifying information that is conducted within their premises can only be done with a court order. It also requires the specific intervention of an officer or employee of the carrier acting in accordance with regulations prescribed by the Federal Communications Commission (FCC).

CALEA (47 USC 1005) directs telecommunications carriers to consult with telecommunications equipment manufacturers to develop equipment necessary to comply with the capability and capacity requirements identified by the FBI. For efficient industry-wide implementation of the above requirements, CALEA (47 USC 1006) directs the law enforcement community to coordinate with the telecommunications industry and state utility commissions to develop suitable technical standards and establish compliance dates for equipment. The FCC may grant extensions to the compliance dates if it determines that the capability requirements are not reasonably achievable within the compliance period.

CALEA (47 USC 1008) gives the Attorney General, subject to the availability of appropriations, authority to pay telecommunications carriers for all reasonable costs directly associated with the modifications performed by carriers in connection with equipment, facilities, and services installed or deployed on or before January 1, 1995 (known as the "grandfather" date).

MAJOR EVENTS FOLLOWING ENACTMENT OF CALEA

Initial Delays

CALEA gave implementation responsibility to the Attorney General, who, in turn, delegated the responsibility to the FBI. The FBI leads that nationwide effort on behalf of federal, state, and local law enforcement agencies. FBI officials initially anticipated that it would take a year for a standard to be developed and agreed upon by law enforcement, the telecommunications carriers, and the equipment manufacturers. Telecommunications consultants estimated that it would take the industry another three years to design, build and deploy new systems to comply with CALEA. Instead, industry and law enforcement became involved in a protracted dispute over what should be required for law enforcement's wiretapping capabilities.

By March 1997, the completion of the capability standard was overdue by 16 months. The FBI attempted to expedite the industry's implementation of CALEA by releasing regulations that included a cost recovery plan for the federal government's payment of costs associated with CALEA, as well as capability and capacity requirements for the industry to meet. The plan required more extensive upgrades to networks than the telecommunications industry believed were necessary for law enforcement to preserve its wiretapping capabilities. Industry groups and privacy advocates disputed the FBI's plan. They argued that the FBI was attempting to expand its surveillance capabilities beyond the congressional intention of CALEA, and was attempting to unfairly shift costs and accountability away from the federal government onto private industry. Furthermore, the industry argued that, without an adopted capability standard, it could not begin designing, manufacturing, and purchasing the equipment to achieve CALEA compliance.

In December 1997, the Telecommunications Industry Association (TIA, representing telecommunications equipment manufacturers) adopted, over the objections of the law enforcement community, a technical standard, J-STD-025, also known as the "J-standard." This standard prescribes upgrades to network devices to meet CALEA's assistance capability requirements for local exchange, cellular, and broadband personal communications services (PCS). Although the FBI claimed that the J-standard did not provide all of the capabilities needed, the industry asserted that CALEA's language stated that telecommunications carriers would be compliant if they met publicly available standards adopted by the industry.

Privacy rights groups, on the other hand, protested two aspects of the J-standard that they asserted would make information beyond what is legally required available to law enforcement. One was a feature enabling the telecommunications network to provide location information for users of mobile wireless telecommunications services. The location information protocols in J-STD-025 allow law enforcement agencies to obtain information on the physical location of the nearest cell site (i.e., the receiver/transmitter antenna and base station) of mobile phone handsets at the beginning and end of each call. Wireless carriers are now deploying another technology (called triangulation) that will enable the carriers, and law enforcement, to track wireless telephone users more precisely, potentially within a few meters. The other was a feature enabling the network to access packet-mode data from telephone calls using more advanced systems. Privacy rights groups argued that these capabilities would violate the Fourth Amendment rights of individuals against unreasonable searches and seizures. Despite these objections, telecommunications manufacturers began designing new switches and upgrades to existing switches according to the J-standard.

The FBI's "Punch List"

In the negotiations in developing the J-standard, TIA had refused to include some of the capabilities that law enforcement officials wanted to facilitate digital wiretapping. As a result, in March 1998, the FBI petitioned the FCC to require the telecommunications industry to adopt those additional capabilities. Industry and privacy rights groups protested that the FBI's plan would unlawfully expand enforcement capabilities. Labeled the "punch-list" by the telecommunications industry (because of its perceived attempt to force the industry to comply), the plan included the capabilities to:

(1) intercept the content of conference calls initiated by the subject under surveillance (including the call content of parties on hold) pursuant to a court order or other legal authorization;

(2) obtain information during conference calls identifying all active parties to the call, and whether a party has been put on hold, has joined, or has been disconnected from the call;

(3) monitor when the subject under surveillance uses "vertical services" such as call-forwarding and call-waiting;

(4) retrieve call timing information, i.e., when calls are initiated and completed;

(5) obtain "dialed digital extraction" capability that shows any digits on the handset pressed by a surveillance subject after a call has been connected, including credit card or bank account numbers;

(6) monitor "surveillance status," requiring carriers to send information to law enforcement agencies to verify that a wiretap had been established;

(7) monitor "continuity check tone," requiring that a dial tone be present on the channel received by law enforcement until the surveillance subject uses the phone;

(8) monitor when a surveillance subject adds or deletes any communications services;

(9) monitor "in-band and out-of-band signaling," requiring carriers to send a notification message to law enforcement when the network sends signals to the subject's phone, such as special rings, busy signals, and call-waiting signals.

Capacity Requirements

The FBI's subsequent implementation actions were also opposed by the telecommunications industry. In March 1998, the FBI announced its estimated capacity requirements for local exchange, cellular, and broadband personal communications services (PCS).[5] The industry protested the FBI's estimates, arguing that it would require telephone carriers to accommodate thousands of wiretaps simultaneously, an impractical and unnecessary burden. In July 1998, the FBI developed guidelines and procedures to facilitate small carrier compliance with its capacity requirements, and asked carriers to identify any systems or services that did not have the capacity to accommodate those requirements. In December 1998, the FBI began a proceeding to develop capacity requirements for services other than local exchange, cellular, and broadband PCS, asked additional questions of interested parties in June 2000.[6] These technologies and services included paging, mobile satellite services, specialized mobile radio, and enhanced specialized mobile radio. To date, that proceeding has not been completed.

FCC Actions

As a result of petitions from the industry and the FBI, the FCC became involved in the implementation of CALEA. In October 1997, the FCC released its first Notice of Proposed Rule Making (NPRM) on CALEA implementation.[7] The NPRM sought comments from interested parties regarding a set of policies and procedures proposed by the FCC for telecommunications carriers to follow. The proposed procedures would (1) preclude the unlawful interception of communications, (2) ensure that authorized interceptions are performed, (3) maintain secure and adequate records of any interceptions, and (4) determine what entities should be subject to these requirements, whether the requirements are reasonable, and whether to grant extensions of time for compliance with the requirements. The telecommunications carriers, privacy rights groups, and the FBI all submitted comments to the FCC to attempt to influence the final decision.

In April 1998, the FCC released a Public Notice requesting comments on issues raised in petitions from industry, the FBI, and privacy rights groups concerning the dates that carriers were required to comply with CALEA and the dispute over the J-standard. Based on comments it received, the FCC extended the deadline until June 30, 2000 for telecommunications carriers to comply with CALEA, stating that without a standard, the necessary equipment would not be available in time.[8]

In October 1998, the FCC initiated a proceeding to review the technical capabilities prescribed by the J-standard.[9] The goal of that proceeding was to determine whether

---

[5] Federal; Register 63, page 12217, FBI, Final Notice of Capacity, March 12, 1998.

[6] Federal Register 63, page 70160, FBI Notice of Inquiry, December 18, 1998, and page Federal Register 65, page 40694, FBI Further Notice of Inquiry, June 30, 2000.

[7] FCC NPRM CC Docket No. 97-213, FCC Record 97-356, released October 10, 1997.

[8] FCC Memorandum Opinion and Order in the Matter of Petition for the Extension of the Compliance Date under Section 107 of CALEA, released September 11, 1998.

[9] FCC Proposes Rules to Meet Technical Requirements of CALEA. Report No. ET 98-8. FCC News, October 22, 1998.

telecommunications carriers should be required under CALEA to meet the FBI's "punch list" items. The FCC addressed these issues in several documents released over the following year. In March 1999, the FCC's First Report and Order established the minimum capability requirements for telecommunications carriers to comply with CALEA.[10] Telecommunications carriers were required to ensure that only lawful wiretaps occur on their premises and that the occurrence of wiretaps is not divulged to anyone other than authorized law enforcement personnel. On August 2, 1999, the FCC decided to allow carriers to decide bow long they would maintain their records of law enforcement's wiretap, pen register, and trap and trace interceptions.[11] On August 31, 1999, the Second Report and Order established a definition for "telecommunications carrier" to include all common carriers, cable operators, electric and other utilities that offer telecommunications services to the public, commercial mobile radio services, and service resellers.[12] The definition did not include Internet service providers (ISPs), which were explicitly excluded under the CALEA statute.

The FCC's Third Report and Order, released August 31,1999, adopted technical requirements for wireline, cellular, and broadband PCS carriers to comply with CALEA requirements.[13] The ruling adopted the J-standard, including the two capabilities that were opposed by the privacy rights groups (i.e., the ability to provide location information and packet-mode data to law enforcement). The FCC also adopted six of the nine punch list capabilities requested by the FBI to be implemented by telecommunications carriers (punch list items 6, 7, and 8 listed above were not granted). The Order required all aspects of the J-standard except for the packet-mode data collection capability to be implemented by June 30, 2000. The Order required carriers to comply with the packet-mode data capability and the six punch list capabilities by September 30, 2001. Some privacy rights groups expressed outrage over the ruling, stating that the FCC had "threatened civil liberties and may even grant the FBI new powers of surveillance." Upon adoption of the ruling, however, FCC Chairman William Kennard stated that "We have carefully balanced law enforcement's needs against the rights of all Americans to privacy, and the cost to industry of providing these tools to assist law enforcement."[14]

ISSUES FOR CONGRESS REGARDING CALEA

Despite the contention among all of the parties involved, issues associated with the implementation of CALEA have not been given a great amount of attention in Congress. While funding for CALEA has been discussed at appropriations hearings, the only oversight hearing focusing specifically on CALEA implementation was by the House Judiciary Committee, Subcommittee on Crime, on October 23, 1997. Furthermore, to date, no reports have been published by the General Accounting Office or the Congressional Budget Office on CALEA implementation issues. However, several issues have not been resolved, and deadlines for implementation have passed while others are approaching.

---

[10] FCC 99-11, Report and Order CC Docket No. 97-213, released March 15, 1999.
[11] FCC 99-184, Order on Reconsideration CC Docket No. 97-213, released August 2, 1999.
[12] FCC 99-229, Second Report and Order, CC Docket No. 97-213, released August 31, 1999.
[13] FCC 99-230, Third Report and Order, CC Docket No. 97-213, released August 31, 1999.
[14] FCC Sides with FBI on Tapping, *Wired News,* August 27, 1999, [http://www.wired.com/news]

Funding

While cost estimates for CALEA implementation continued to rise after its enactment, funding was slow to come. CALEA authorized $500 million to be distributed to telecommunications carriers for implementation. Shortly after enactment, the U.S. Telecommunications Association (USTA), representing over 1,200 large and small telephone companies called the local exchange carriers (LECs), estimated the cost at $2 billion. By 1999, some industry estimates placed CALEA costs at $5 billion. The Department of Justice (DOJ) has disputed the higher estimate. However, the Attorney General did acknowledge that "in excess of $2 billion would be needed by the government to reimburse telecommunications carriers to cover the costs of modifying this enhanced embedded base."[15]

No funding, however, was appropriated for CALEA for three years after enactment. The Omnibus Consolidated Appropriations Act of 1997 (P.L. 104-208) created the Telecommunications Carrier Compliance Fund (TCCF) for CALEA implementation. The 1997 Act provided $60 million to DOJ for CALEA implementation, which could not be released until the implementation plan was approved by House and Senate Appropriations Committees. The 1997 Act also authorized agencies with law enforcement and intelligence responsibilities to transfer unobligated balances into the TCCF.

The FY1998 Appropriations Act for Commerce, Justice, State Departments and Related Agencies (P.L. 105-119), did not provide additional funding to the TCCF. The accompanying conference report (H. Rept.105-405) stated that there was already $101 million in the Fund "which is sufficient to support reimbursement to the telecommunications industry during FY1998."[16] The FY1999 Omnibus Appropriations Act (P.L.105-277) also did not provide any new funding, although a small amount of interest accrued to the Fund, and transfers were made from other federal agency accounts, bringing the total to $102.5 million. DOJ received an additional $15 million in FY2000 appropriations for CALEA implementation.

For FY2001, the Administration's budget request included $120 million to be distributed by DOJ and another $120 million to be distributed by the Department of Defense (DOD) for CALEA implementation.[17] The FY2001 Commerce, Justice, State Appropriations Act (P.L. 106-553) provided another $201 million for the TCCF.[18] In addition, a large portion of the TCCF appropriations was inserted into the FY2001 Military Construction Appropriations Act (P.L. 106-246). This law was enacted on July 13, 2000, under a section providing supplemental appropriations for FY2000 "for an additional amount for 'Salaries and Expenses', $181 million to remain available until expended, which shall be deposited in the TCCF," under the Drug Enforcement Administration supplemental appropriations. In total, Congress has appropriated $499,557,270 for the TCCF. As of January 15, 2001, the unobligated balance of the TCCF is $227,757,270, although no money has yet been provided

---

[15] Letter from Attorney General Janet Reno to Honorable Ted Stevens, Chairman of Senate Committee on Appropriations, October 6, 1998.

[16] Telecommunications Carrier Compliance Fund, *Congressional Record,* November 13,1997, page H10836.

[17] Budget of the United States Government FY2001, Appendix, page 631, submitted February 7, 2000. The rationale behind providing some of the funds to DOD, to be redistributed to DOJ, was that law enforcement often assists DOD in conducting wiretaps on suspected international criminals, in which case the FBI's CALEA capabilities could be used for national security purposes.

[18] All of this amount is provided under DOJ appropriations even though $141.3 million of it is designated for national security purposes.

to the carriers. The Administration is expected to request significant additional TCCF funding for FY2002.

In addition, the FBI's costs for CALEA implementation have been criticized by telecommunications industry groups as being exorbitant. The FY2001 Commerce, Justice, State Appropriations Act (P.L. 106-553) provides $17.3 million for the FBI's CALEA implementation activities. FY2000 appropriations for the FBI's CALEA implementation was $15 million to remain available until expended. The FBI declined to provide any information on its funding or expenditures for CALEA implementation.

Equipment and Standards

The industry standard (J-STD-025) with added "punch list" items adopted by the FCC in August 1999, defined services and features to support electronic surveillance by law enforcement agencies. The J-standard includes interfaces to provide law enforcement agencies access to packet-mode data and location information. Privacy rights groups criticize this provision, claiming that it essentially creates a situation in which mobile telephones can be converted into location-tracking devices. They argue that neither the packet-mode data, nor the location information, are authorized by CALEA, and that these capabilities violate the Fourth Amendment rights of individuals against unreasonable searches and seizures.

Both the DQJ and the FBI expressed satisfaction with the FCC's requirements for carriers, even though the FCC did not include three of the FBI's "punch list" items in its decision. Attorney General Janet Reno stated that "This ruling will enable law enforcement to keep pace with these changes [in telecommunications technology] and ensure we will be able to maintain our capability to conduct court-authorized electronic surveillance."[19] Some observers questioned whether the three deleted punch list" items were ever necessary for law enforcement's crime fighting purposes. Privacy rights groups questioned whether any of the punch list capabilities were necessary for law enforcement to conduct wiretaps in the future.

Many carriers were not able to meet the FCC's June 30, 2000 deadline for upgrading their systems to be compliant with the J-STD-025 (required for all carriers by the FCC's Third Report and Order). Of the approximately 3,600 telecommunications carriers in the Untied States (including wireless, wireline, cable, and international carriers), about 2,000 of them did not meet the deadline. Many of them petitioned the FCC for an extension, arguing that CALEA-compliant equipment and software had not become available as extensively as had been expected.[20] The FCC granted an extension to some petitioners until March 31, 2001, and is reviewing other petitions on a case by case basis. In the majority of petitions, carriers claim they will become CALEA-compliant by 2002.

The next CALEA deadline for carriers to meet is March 12, 2001, as defined by the FBI (63 Fed. Reg. 12217, March 12, 1998), when carriers providing local exchange, cellular services, and broadband PCS must comply with the capacity requirements. This entails additional hardware and software upgrades for all switches operated by carriers to enable (multiple, in some cases) electronic surveillance actions at any given time. The FBI

---

[19] Justice Department press release regarding the FCC's CALEA standards, August 27, 1999

[20] FCC *Public Notice*, rpt. No. CALEA-001, CALEA Section 107(c) Extension Petitions Filed, released June 30, 2000.

determined a separate capacity requirement for each county in the United States. The telecommunications industry disputed the number of wiretaps the FBI required carriers to support, arguing that law enforcement has never before asked for a wiretap at many of the locations indicated. The FCC decided that capacity requirements fall outside of its jurisdiction, and has not become involved in the dispute between carriers and the FBI on that issue. Many carriers have petitioned the FBI for an extension of that deadline, based on the lack of availability of equipment and the lack of resources to make the upgrades.

By September 30, 2001, wireline, cellular, and broadband PCS carriers are required to implement the additional "punch list" capabilities adopted by the FCC. On August 23, 2000, the Cellular Telecommunications Industry Association filed a petition for a waiver of that deadline for its cellular and PCS provider members.[21] Other carriers may also petition for an extension to that deadline. On September 29, 2000, TIA submitted a report to the FCC on the technology and privacy issues involved in applying CALEA to packet-mode services.[22] As a result of that study, TIA recommended that the FCC to suspend the compliance date for packet-mode communications requirements, based on the uncertainty for equipment manufacturers and carriers in knowing how to satisfy CALEA requirements following a recent court decision (Privacy Issues Associated with CALEA, below) that questioned the legal requirements for law enforcement agencies to monitor packet-mode communications.

Impact on the Telecommunications Industry

The FBI estimated that about 3,600 telecommunications companies defined as "carriers" by the FCC's Second Report and Order are required to upgrade their systems for CALEA-compliance. These include local exchange carriers, interexchange carriers, competitive access providers (also called service resellers), wireless service providers, satellite communications companies, and any other company that offers telecommunications services to the public. Since few cable television companies, electric and other utilities currently offer telecommunications services to the public, those industries are not significantly affected.

To facilitate CALEA implementation, in March 2000 the FBI established a "Flexible Deployment Assistance" program, through which carriers could file for an extension to a CALEA deadline.[23] To be considered for an extension by the FCC, carriers are required to submit detailed information to the FBI on the types of equipment they had deployed.[24] Because the FCC made this information a prerequisite to the granting of an extension, the FBI collected the information primarily from carriers that filed for an extension.[25] The FBI plans to determine which equipment within the higher priority jurisdictions (identified by state and local law enforcement agencies) must have CALEA upgrades installed first. Only the

---

[21] For a copy of the filing, see [http://www.wow-com.com/lawpol/filing/pdf/ctia082300.pdf]

[22] Report to the FCC on Surveillance of Packet-mode Technologies, September 29, 2000.
    http://www.tiaonline.org/government/filings/JEM_Rpt_Final_092900.pdf.

[23] The FBI's "Flexible Deployment Assistance Guide" containing questions for carriers to answer in order to file for an extension to a CALEA deadline, can be found at CTIA's website at [http://www.wow—com.com/lawpol/guide.cfm].

[24] The FCC has regulatory authority to grant extensions to telecommunications carriers, but under CALEA, it must consult with the Attorney General before granting an extension.

[25] Some carriers unknowingly provided the information requested in the Flexible Deployment Assistance Guide and sent it to the FBI without needing to file for an extension.

equipment in the lower priority areas will be granted extensions. It is not clear when the FBI wilt release its list of priority areas.

Under the CALEA statute, carriers will not be reimbursed for CALEA upgrades to network devices that were deployed or installed after January 1, 1995. All of the PCS infrastructure (which was deployed after January 1995), and most of the infrastructure of the cellular industry (some of which was in place prior to 1995) is exempt from reimbursement under CALEA. CALEA further states that equipment installed prior to January 1, 1995 does not have to be upgraded unless the FBI decides that it is necessary and pays all of the costs of the upgrade. Carriers will be responsible, however, to make and pay for CALEA upgrades to equipment that has had "significant" upgrades since January 1995. Most equipment in the wireline telephone infrastructure has bad some level of upgrades (e.g., corrections for the year 2000 computer problem). That fact has caused many in industry to speculate that much of the costs for CALEA upgrades on equipment deployed even before 1995 will have to be paid by industry.[26] Many other commercial mobile radio services such as enhanced paging, specialized mobile radio, multichannel multipoint distribution service, and local multipoint distribution service, must also make and pay for upgrades to their systems.

Any costs not reimbursed by the TCCF could ultimately be passed on to consumers. Some argue that CALEA has created an incentive for telecommunications carriers not to make any general upgrades to their networks due to concerns that the FBI will deem the upgrades to be "significant" and deny reimbursement for CALEA upgrades. Also, some claim that as a result of FBI interference with the business plans of telecommunications firms, U.S. companies have been hindered in their business transactions with foreign companies.

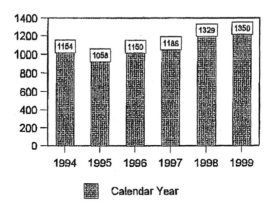

Figure 1: Total Federal and State Wiretap Authorizations

Source: Report of the Director of the Administrative Office of the U.S. Courts on Applications for Orders Authorizing Interception of Wire, Oral, or Electronic Communications, published April 2000

Given that CALEA has still not been implemented over six years after enactment, some question whether it was ever necessary. Privacy rights groups argue that there is insufficient evidence that law enforcement has been hampered without the new capabilities to be provided by CALEA. As shown in Figure 1, the number of authorized state and federal wiretaps has

---

[26] According to one estimate, 30-50% of the wireline industry will have to make and pay for CALEA upgrades.

increased an average of three percent per year since CALEA's enactment, to a total of 1,350 such actions in 1999. During this period, all but three requests for wiretaps were authorized by the courts. Surveillance of new technologies (such as those provided by wireless or other advanced services) accounted for over half of all requests. Privacy rights groups point to that data to argue that the advent of digital telephone networks has not impeded law enforcement from performing wiretaps, counter to what has been argued by the FBI. Others might argue, however, that crimes have gone undetected as a result of the lack of CALEA upgrades.

Some question whether CALEA can ultimately solve law enforcements problem of being able to perform wiretaps in a digital telecommunications environment. Given the extensive delays in implementing CALEA and the extensions granted to many telecommunications carriers, some question whether the industry, law enforcement, and privacy rights differences will ever be resolved, or whether the FCC's decisions will ever be upheld by the courts. Even if all telecommunications carriers finally become fully CALEA-compliant, some believe that in the future, strong encryption will be used often enough in telecommunications (both voice and data) that law enforcement will be unable to decipher the content of communications that it collects.[27] Some question whether the goal of CALEA to preserve law enforcement's wiretapping capabilities in the digital age, equal to those capabilities under traditional analog telecommunications systems, is still realistic.

Privacy Issues Associated with CALEA

Privacy rights groups are concerned that the FBI, in its implementation actions, is trying to usurp greater capabilities than were intended by CALEA. They claim that the FCC's interpretation of CALEA has endangered the public's right to privacy by permitting law enforcement to obtain the actual content of conversations over telecommunications networks, instead of only the call origination information (e.g., the number dialed), in violation of the Fourth Amendment.

Some telecommunications industry groups claim that the FBI's Flexible Deployment Assistance program violated the privacy rights of telecommunications carriers by requiring them to submit detailed information on their networks in order to be considered for an extension to the June 30, 2000, deadline of CALEA implementation. The FCC did not provide guidance to carriers until April 25 on how to file for extensions to the June 30 deadline, thus, carriers contended, giving them little time to dispute the FBI's procedures.[28] Telecommunications industry representatives argue that the FBI can now use the proprietary information provided by the carriers to deny reimbursement for future CALEA upgrades.

In January 2000, some privacy rights groups, together with telecommunications industry groups, filed suit with a federal appeals court to block the FBI's requirements placed on carriers and on mobile telecommunications devices.[29] The lawsuit was based on arguments over privacy rights, disputes over the equipment standards adopted by the FCC, the FBI's

---

[27] Will Crypto Feast on Carnivore? *Wired News*, August 4, 2000
    [http://www.wired.com/news/0,1294,37915,00.html].
[28] FCC *Public Notice* 00-154, CALEA Section 103 Compliance and Section 107(c) Petitions. April 25, 2000.
[29] Three law suits were originally filed with the U.S. Court of Appeals for the DC Circuit against the FCC's August 1999 rulings. The Court consolidated the three petitions into a single petition which can be found at [http://www.epic.org/privacy/wiretap/].

capacity requirements, and the FBI's procedures for making payments to the carriers for upgrades. On August 15, 2000, the Court upheld some aspects of the J-standard, while overturning other aspects.[30] The court decided that the FCC correctly required carriers to build into their systems a capability to provide location information on wireless phones, but that the requirement is limited to the antenna handling the call at the time of interception. The court also ruled that government agents must meet the highest legal standards if they want to intercept packet-mode data from telephone signal transmissions. The court further ruled that the FCC exceeded its statutory authority granted by CALEA in its adoption of four of the six punch list capabilities adopted by the FCC, and remanded consideration of those requirements back to the FCC.[31] The FCC is currently considering alternatives for how to satisfy the court's concerns.[32]

THE INTERNET DIMENSION: THE FBI'S CARNIVORE SYSTEM

Since personal and business communications are being conducted increasingly over the Internet (in both e-mail and Worldwide Web transactions), the law enforcement community has become interested in being able to conduct wiretapping over that medium. In 1999, the FBI asked the Internet Engineering Task Force (IETF), an industry-led standards-setting body for Internet protocols, to consider adopting protocols for wiretapping for use by Internet service providers. Several individuals and groups protested this suggestion. They argued that this development would harm network security, result in an increase in illegal activities, diminish users' privacy, stifle innovation, and impose significant costs on developers of communications equipment and services. In November 1999, the IETF decided overwhelmingly to reject the FBI's proposal.

On February 28, 2000, at a joint hearing of the House Judiciary Crime Subcommittee and the Senate Judiciary Oversight Subcommittee, the DOJ and the FBI argued that having the authority to conduct wiretaps over the Internet would help law enforcement in investigating cyber-attacks. Unbeknown to the public (or to the White House, according to some reports), however, the FBI had already developed a special software program, called "Carnivore," to efficiently collect Internet communications. The Carnivore system includes a "network analyzer" or "sniffer" that collects information from e-mail and other Internet communications.[33] A separate device that runs the Carnivore program is installed at each Internet Service Provider (ISP) to sort through the incoming and outgoing traffic at the ISP to find information flowing to and from a person under investigation. Prior to activating these systems, the FBI obtains court orders for a pen register/trap and trace surveillance. The FBI does not use Carnivore for many of the larger ISPs, which maintain in-house capabilities to

---

[30] U.S. Telecom Association et al., Petitioners v. Federal Communications Commission and United States of America, Respondents AirTouch Communications Inc., et al., Intervenors.
    See [http://pacer.cadc.uscourts.gov/common/opionions/200008/99-1442a.txt] for the court ruling.
[31] The four punch list capabilities that were remanded to the FCC are items 2, 4, 5, and 9 in the list provided above.
[32] FCC *Public Notice* DA 00-2342, Commission Seeks Comments to Update the Record in the CALEA Technical Capabilities Proceeding CC Docket No. 97-213, October 17, 2000
[33] A sniffer is a common program used by network administrators to analyze the flow of communications "traffic," detect bottlenecks and keep traffic flowing efficiently.

enable law enforcement's Internet monitoring. There are, however, probably thousands of ISPs that do not have that capability.

The existence of Carnivore was discussed at a hearing on July 24, 2000, held by the House Judiciary Committee, Subcommittee on the Constitution. Some witnesses at the hearing were concerned that the Carnivore system could enable the collection of the content of communications, equivalent to a wiretap (which requires a higher burden for law enforcement to show that a crime is likely being committed). The FBI claimed that only information or messages of relevance to a court-approved criminal investigation are stored and reviewed, and other information is discarded. Privacy and civil rights groups question whether anything prevents other non-approved communications from being collected.

Privacy Issues Associated with Internet Monitoring

Many have seen a parallel between the privacy issues raised in the implementation of CALEA and those associated with monitoring Internet communications. Some suspect that the FBI is trying to avoid the pitfalls it encountered in implementing CALEA in its efforts to intercept Internet traffic. They argue that the FBI may be trying to convince the computer industry to adopt a single standard for Internet monitoring technology. Then, as Internet communications grow, the FBI could monitor Internet traffic more efficiently. On April 6, 2000, the House Judiciary Committee, Subcommittee on the Constitution, held a hearing on the Fourth Amendment and the Internet. Several witnesses at the hearing warned of the lack of legal protections for the privacy of Internet communications, while one witness believed that the existing laws governing privacy on the Internet were adequate.[34]

It is technically possible for the FBI to use systems such as Carnivore to collect the content of communications of individuals who have not been served a court order, in violation of the Fourth Amendment. Some call for placing a greater burden on the ISPs to protect the privacy of communications to which government access has not been granted. Others argue that once a system such as Carnivore is installed, the ISP has little control over what information the FBI accesses, and therefore should be immune from prosecution. Legal issues related to Internet monitoring are becoming increasingly urgent as the technical tools for online surveillance (and the ability to thwart those activities) improve and proliferate.

In addition, two relatively new services might present privacy rights issues in connection with CALEA implementation and Internet monitoring. These services are Internet access over mobile telephones and the placement of phone calls via the Internet.[35] Each of these services raises the question of whether they should be regulated as Internet or telephone communications, and over what privacy protections they should be granted. As various communications technologies (including the telephone, the internet, digital television, and various wireless systems) continue to converge, the application of regulations and laws governing their separate uses has become an issue of potential congressional interest.

The Attorney General agreed to conduct an in-house review of Carnivore, but declined to release the system's source code or design to the public, citing the potential for revealing its

---

[34] Testimony from the House Judiciary Committee, Subcommittee on the Constitution, hearing April 6, 2000.

weaknesses. On August 2, 2000, a federal judge ordered the FBI to expedite its review of a request made by privacy rights groups for the immediate release of details on the technology and capabilities of Carnivore. On August 16, in response to a court order the FBI announced that it would review 3,000 pages of documents pertaining to Carnivore to determine what can be released under the Freedom of Information Act (FOIA), and release information in increments every 45 days. Privacy rights groups objected to the FBI's caveat that some of the information might be redacted (i.e., edited or revised prior to release) based on FOIA exemptions covering national security, privacy, and other concerns. On October 2, 2000, the FBI released the first set of documents from its files on Carnivore, most of which was redacted or completely withheld.

On September 6, 2000, the Senate Judiciary Committee held a hearing on the Carnivore controversy. At the hearing, FBI officials testified that over the past two years they had used the system about 25 times, and only with the permission of the courts. A witness from the Center for Democracy and Technology, a privacy rights group supported in part by the telecommunications industry, argued that the potential for government abuse is high and suggested that the Internet Service Providers should control the operations of the Carnivore system. Senator Hatch, the Committee Chairman, and another witness, Dr. Vinton Cerf, founding president of the Internet Society, raised questions about whether private sector corporations can be trusted any more than the government.[36]

In response to the congressional inquiries and media reports, the Attorney General agreed to an "independent technical review" of Carnivore. Several leading academic institutions declined to bid on the DOJ's proposal, however, due to concerns that the DOJ reserved the right to edit or omit sections of the report.[37] On September 26, the DOJ selected the Illinois Institute of Technology Research Institute (JITRI) to review the system. Its report, released on November 21, concluded that Carnivore functions as the FBI had described, and generally does not "over-collect" information. The report also stated, however, that "While the system was designed to, and can, perform fine-tuned searches, it is also capable of broad sweeps. Incorrectly configured, Carnivore can record any traffic it monitors." The report further stated that "LITRI did not find adequate provisions (e.g., audit trails) for establishing individual accountability for actions taken during use of Carnivore."[38] The report also made recommendations for improving Carnivore for efficiency and protecting the privacy of Internet users.

On November 21, 2000, the Chairman and Ranking Member of the Senate Judiciary Committee asked the FBI to "explain why Carnivore was tested to determine if it was capable of intercepting and archiving unfiltered traffic through an ISP, whether Carnivore in fact has that capability, and under what circumstances it could ever be legitimately used to draw on that capability." They also requested "complete and unredacted copies of the documents produced in response to the FOIA lawsuit together with any other documents related to Carnivore's capability to intercept and archive unfiltered traffic."

---

[35] Internet telephony, also called voice over Internet protocol (VOIP), is the transmission of voice traffic as data packets over a packet-switched data network instead of as a synchronous stream of binary data over a conventional circuit-switched voice network.
[36] Senate Judiciary Committee hearing, September 6. 2000. Federal News Service Transcript.
[37] "Universities Unwilling to Review FBI's 'Carnivore' System," [http://www.CNN.coml2000/TECH/computing/09/06/camnivore/index.html].
[38] IITR1 Independent Technical Review of the Carnivore System, Draft Report, 17 November, 2000. [http://www.usdoj.gov/jmd/publications/carniv_entry.htm].

Given the increase in Internet communications and the increase in law enforcement's electronic surveillance activities, the FBI is likely to increase its use of Carnivore systems in its Internet monitoring efforts. In addition, the September 2000 TIA report to the FCC on surveillance of packet-mode technologies, suggested using Carnivore technology for telephone wiretaps as a way of meeting CALEA requirements and maintaining a separation between content and packet information.[39] Many industry observers argue that the merging of technologies used for telephone and Internet communications has necessitated a change in law to establish a parity of procedures for digital surveillance by law enforcement.

RELEVANT LEGISLATION IN THE 106TH CONGRESS

Several bills were introduced in the 106th Congress concerning electronic surveillance. On April 21, 1999, the Electronic Rights in the 21st Century Act was introduced (S. 854, Leahy). The bill is intended to protect the privacy of the electronic communications of individuals by increasing the requirements for a warrant or subpoena before the government could obtain the electronic communications contents or location information from a service provider, among other provisions. While not affecting CALEA implementation directly, the bill would increase the burden on law enforcement in obtaining a court order, and could reduce the number of wiretaps performed. The bill was referred to the Senate Judiciary Committee and no further action was taken.

The Notice of Electronic Monitoring Act (H.R. 4908, Canady, introduced July 20, 2000) would have required employers to notify their employees when they monitor e-mail communications and other computer usage in the workplace (companion bill, S. 2898, Shumer, introduced July 20). The bills were referred to the House and Senate Judiciary Committees, respectively, with no further action taken.

Two bills intended to limit law enforcement's electronic surveillance activities were introduced on July 27, 2000. The Digital Privacy Act of 2000 (H.R. 4987, Barr) would have increased the reporting requirements for the Attorney General on the electronic surveillance activities of the DOJ. The bill would also increase limits on government access to contents of stored electronic communications (including e-mail) and to location information of mobile electronic devices. The Electronic Communications Privacy Act of 2000 (H.R. 5018, Representative Canady) would also increase the Attorney General's reporting requirements on the electronic surveillance activities of the DOJ, and excludes electronic communications (both stored and realtime), if seized without proper authority, from being used as evidence in court or agency hearings. H.R. 5018 originally did not include a section on limiting government access to content of stored electronic communications (later added in an amendment), and includes other specific differences from H.R. 4987.[40]

The House Judiciary Committee, Subcommittee on the Constitution, held a hearing on September 6, 2000, on H.R. 4908, H.R. 4987, and H.R. 5018. Privacy rights groups and industry representatives supported most aspects of the three bills. The DOT expressed concern that this legislation could interfere with law enforcement's ability to conduct digital surveillance. On September 14, H.R. 5018 was adopted by the Subcommittee on the

---

[39] *FCC Could Adopt Carnivore,* Wired News, September 29, 2000.
[40] For a detailed description of H.R. 5018, see CRS Report RS 20693 by Gina Marie Stevens.

Constitution, with an amendment added from part of H.R. 4987 to limit government access to location information of mobile electronic devices. The House Judiciary Committee approved H.R. 5018 (amended) on September 26 (H.Rept. 106-932 released October 4, 2000). No further action was taken on either of these bills.

Chapter 9

COMMERCIAL REMOTE SENSING BY SATELLITE: STATUS AND ISSUES

Richard E. Rowberg

INTRODUCTION

Background

Since the beginning of the space age, observation of the Earth (remote sensing) by satellite and human-occupied space vehicles has played an important role in U.S. and international space programs. The earliest remote sensing satellites were used for national security purposes.[1] Early on, however, space policy makers and others recognized the potential importance of remote sensing for civilian purposes such as environmental and climate monitoring. Such satellites began appearing in the early 1960s. Furthermore, once these satellites were in operation, many in the space community began advocating transferring responsibility for developing and operating civilian remote sensing satellites to the private sector. This effort became the policy of the United States in 1979, and, since then, congressional and administrative action has attempted to implement that policy. The Land Remote-Sensing Commercialization Act of 1984 (P.L.98-965) Act of 1984 set out terms for transferring the government owned Landsat satellite program to the private sector. The Land Remote Sensing Policy Act of 1992 (P.L. 102-555) declared commercialization of land remote sensing to be a long-term policy goal of the United States and established procedures for licensing private remote sensing operators.[2]

While a commercial satellite remote sensing industry has emerged in the United States in recent years, it has not been the success envisaged by its early proponents. Competition from aerial remote sensing (aircraft, balloons), the slow development of a market for remote sensing products outside local, state, and federal governments, competition from government-subsidized, foreign remote sensing satellites, and regulations resulting from national security

---

[1] Remote sensing satellites for national security purposes are called reconnaissance satellites.
[2] P.L. 102-555, sec. 2 and sec. 201.

concerns, among other factors, have slowed the development of a heathy commercial satellite remote sensing industry. A commercial remote sensing industry has been established, however, and by all accounts is growing.[3] The images of the aftermath of the terrorist attack on the World Trade Center obtained by the lkonos 2 satellite operated by Space Imaging gave particular prominence to the existence of the commercial remote sensing industry in this country.

While the industry is growing, however, its future appears uncertain. The industry has not turned a profit and still depends largely on purchases of images by the federal government to remain in operation.[4] While action has been modest in the 107th Congress to date, congressional interest remains strong in the development of a healthy commercial remote sensing industry.[5] This report presents a review of the current status of the industry and an analysis of key issues that are likely to affect its future. It begins with a brief description of the technology of satellite remote sensing followed by a brief review of the history of its development. Next, the report examines three areas that encompass most of the important issues relating to the industry's development. They are: the commercial market potential and the role of the federal government as a purchaser and provider of remote sensing images; national security concerns about commercial remote sensing satellites; and federal regulation of the industry. Finally, the report presents a discussion of policy considerations.

SATELLITE REMOTE SENSING TECHNOLOGY, APPLICATIONS, AND HISTORY

Technology

Remote sensing is defined as any observation made at a point removed from the object under investigation. More commonly it refers to observations of areas of land and water covering the earth by airplane or satellite. Aerial remote sensing currently operates from altitudes of about 500 meters to 20 kilometers (kin) with the latter being the domain of aircraft such as the U2 for national security purposes. The range of operation for most spacecraft and satellite remote sensing is about 250 km to 1000 km with satellites close to the 1000 km level.[6] Some remote sensing satellites also operate in geostationary orbit at 36,000 km. By the nature of that orbit, these satellites remain over the same place on Earth at all times.

Remote sensing involves collecting an image of a region on the Earth by one of two means: passive or active.[7] Passive sensing, which is the mode of operation of most remote sensing satellites today, monitors the objects under investigation by using electro-optic sensors to collect solar radiation reflected off the object. Active sensing uses a source of

---

[3] Barnaby J. Feder, "Bird's-Eye Views, Made to Order," *New York Times*, October 11, 2001, sec. F1, 1.
[4] John C. Baker, Kevin M. O'Connell, and Ray A. Williamson. "Conclusions" in *Commercial Observation Satellites*, ed. by John c. Baker, et al. (Arlington, VA and Bethesda, MD: RAND and ASPRS, 2000) 559.
[5] To date, one bill affecting the commercial remote sensing industry has been introduced. That is H.R. 2426, the Remote Sensing Applications Act of 2001.
[6] Paul J. Gibson, Introductory Remote Sensing: Principles and Concepts (London: Routledge, 2000), 2.
[7] Most of the information in this section is based on material found in Gibson, *Introductory Remote Sensing*, 12-126.

electromagnetic radiation — usually radar — carried on board the satellite. These types of systems are discussed below.

The span of electromagnetic radiation (spectrum) from the sun used by passive remote sensing ranges from the familiar visible radiation to the near and thermal infrared. Not all of this spectrum will pass through the earth's atmosphere, however, so sensors on a remote sensing system have to be designed to receive those portions that can penetrate the atmosphere. Also, not all of the radiation monitored by a remote sensing satellite comes from reflections off the Earth's surface. Radiation is reflected off clouds and other particles in the atmosphere thereby giving these satellites the ability to monitor phenomena such as temperature and water vapor content at different atmospheric layers. As for reflection off the surface, the nature of the object from which the radiation reflects determines where in the spectrum it lies. As a result, different surface features such as vegetation, water, rocks, and populated areas have different spectral signatures. In addition, other factors such as the relative position of the sun and surface coverings such as snow affect that signature. Because of this behavior, remote-sensing systems can be designed for detailed study of many aspects and characteristics of the planet and its atmosphere.

Remote sensing sensors are generally classified as panchromatic (PAN) or multispectral (MS). The former covers the visible spectrum while the latter consists of a series of sensors each set to record simultaneously a different color or portion of the spectrum. MS sensors record images that provide the means to identify and study characteristics of different surface features. These sensors usually include portions of the spectrum in both the visible and infrared regimes. Recently, hyperspectral sensors have been developed which divide up the spectrum into many more, smaller bands than MS sensors for even more detailed characterization of surface features.

The first remote sensing satellites recorded images on film. All current satellites use digital systems to record the images, and have many advantages over film systems. In particular, digital image data can be transmitted to ground-based stations and digital data are able to be processed on computers.

An important feature of remote sensing systems is resolution. There are four measures of resolution: spectral resolution which is measure of the narrowness of the spectral band that can be determined; temporal resolution which measures the frequency at which data of the same region can be obtained; radiometric resolution which measures how many levels of gray can be determined on a black and white image; and spatial resolution. The latter gives the smallest dimension an object can have and still be distinguished from other objects. It is the measure usually cited when characterizing a remote sensing satellite. The Ikonos 2 satellite has a spatial resolution of 0.82 meters for PAN images and four meters for MS images. [8] Spatial resolution is primarily dependent on the optical system used to collect the reflected radiation.

Temporal resolution or revisit time is an important feature of remote sensing satellites along with the area of ground that can be observed at any given instant. The lower the orbit, the shorter the revisit period. Lower orbits, however, also reduce the area of ground that can be covered. By using sensors that can look both forward and back from the satellite, a particular point on the surface can be observed much more frequently than by the revisit rate.

---

[8] Frank Sietzen, Jr., "Advanced imagery raises the ante," *Aerospace America*, Sept. 2001, 39.

The other type of remote sensing, active systems, primarily use radar as the source of electromagnetic radiation. Radar has two major advantages over reflected solar radiation: it can penetrate cloud cover and it can be used at night because it does not depend on the sun. Active systems, therefore, can make observations that cannot be made by passive systems. Radar observations, however, are more complex and costly than passive systems and have been slower to develop. Advances in radar imaging technology, however, have increased interest in these systems in recent years. In particular, synthetic aperture radar (SAR) greatly increases resolution along the path of the satellite (azimuthal resolution) by increasing the effective size of the radar antenna. The effective increase is well beyond the practical limit of a conventional antenna. The improvement is so great that all recent and proposed radar remote sensing satellites are using or will use SAR. One application of SAR is the use of radar to determine variations in surface elevation.

In addition to technologies for obtaining images, a very important aspect of remote sensing is the processing of the raw images to facilitate the analysis and application of the data. This so-called value-added step consists primarily of the development and application of software to analyze the content of the images. Because the latter arc in a digital format, much of this processing is done with the use of computers. One example is the combination of raw images with geographic information systems (GIS) to enable the updating of GIS databases with remote sensing data.[9] In addition to computer software, human expertise in interpreting remote sensing images is an important component of this value-added step.

Applications

Applications of satellite remote sensing are varied and growing.[10] The type of application varies primarily with the portion of the spectrum that can be observed by a sensor, by the spatial resolution of that sensor, and by whether the sensors are passive or use radar. Most current satellites carry a range of sensors so that they can perform a variety of applications. A small but growing number of remote sensing satellites, however are single purpose systems.

Satellites with low spatial resolution — over one kilometer (kin) — are generally used for collecting meteorological and environmental data. An example is the NOAA meteorological satellite series operating in an orbit of 855 km using a multispectral sensor with a resolution of 1.1 km. Meteorological data include monitoring cloud cover, climate change and the atmosphere, rainfall, storm events, and the oceans. The hole in the ozone layer over Antarctica was first observed by the Nimbus 7 satellite in the early 1970s. Environmental observations include vegetation assessment, monitoring of natural hazards (primarily major weather events), and geology. The last item includes large area mapping including entire continents, and the use of thermal infrared data for monitoring volcanoes.

Medium resolution satellites (less than 100 m) are used primarily for environmental monitoring and a range of other applications including military observations and mapping of urban areas. An example of these satellites is the Lands at series now operated by the National Aeronautics and Space Administration (NASA) and the United States Geological

---

[9] Kevin O'Connell and Beth E. Lachman, "From Space Imagery to Information: Commercial Remote Sensing Market Factors and Trends," *Commercial Observation Satellites, 64.*

[10] Much of this section is largely based on material found in Gibson, *Introductory Remote Sensing,* 127-150.

Survey (USGS) of the Department of Commerce. The most recent — Landsat 7 — operates in an orbit of 705 km using a 15 m panchromatic (PAN) sensor and a 30 m multispectral (MS) sensor. Landsat 5, which is still operating, has a 30 m PAN sensor and an 80 m MS sensor. Environmental applications include monitoring ocean and other marine conditions such as thermal pollution, mapping of vegetation properties such as crop health, and the mapping of geological features such as faults. It was a Landsat satellite that detected a thermal plume in Chernobyl Lake, giving the world the first notice of the 1986 accident at the Soviet nuclear reactor. Because the revisit time of these satellites is relatively long— several days — their ability to study the effects of pollution is limited. In another category, Landsat and the French satellite SPOT were used during the Gulf War to monitor oil field fires in Kuwait and other events of military interest.

High resolution remote sensing satellites (10 m and less) are relatively recent. The Ikonos 2 satellite, owned and operated by the U.S. company Space Imaging, was launched in 1999 and has a 0.82 m PAN sensor and a 4 m MS sensor. Recently, India launched a 1 m PAN satellite and several other countries have operating or are planning launches of high resolution remote sensing satellites.[11] A range of mapping applications are possible with these satellites including city planning, transportation and development planning, property marketing, and pollution monitoring. These satellites also offer increased potential for national security applications. Ikonos 2 is being used by the U.S. military in the current war in Afghanistan. In addition, images from that satellite were used extensively by the media following the September 11 attacks on the World Trade Center and the Pentagon. A range of other national security applications, including force monitoring and treaty verification, are envisaged for these high resolution satellites.

The final category is radar imaging. This category is currently represented by a few satellites using synthetic aperture radar (SAR) sensors, including Canada's RADARSAT-1 with an 8 m SAR system. The special characteristics of radar sensing make it useful for meteorological, environmental, and mapping applications. In the case of weather, wind speed and ocean wave height can be measured. Images of surface texture that can be obtained using radar make possible observations of oil slicks, land and sea ice, and ocean currents, among other things. Topological measurements are also possible because of the ability of radar sensing to determine changes in surface height. Information on deforestation, crop growth, and hydrological processes, particularly in regions with extensive cloud cover, can be obtained with radar systems.

Applications of remote sensing are many and varied. With the growth of high-resolution sensors, they should expand into areas currently the purview of aerial remote sensing. How competitive satellite remote sensing will be and whether it can capture a significant portion of the commercial market, such as mineral and petroleum exploration and real estate development, remains to be seen. This issue will be discussed more fully below.

---

[11] See Frank Sietzen, Jr., *Aerospace America,* 39, for a list.

History

Remote sensing by satellite began in 1960 with the CORONA project, a series of high-resolution satellites used for surveillance of the Soviet Union. These satellites operated at low orbits (150-450 km) and used cameras to obtain photographic images.[12] Also in 1960, the United States launched the first weather satellite, TIROS-1. Its purpose was to provide early warning of major storms. In 1972, the first Earth monitoring satellite — Earth Resources Technology Satellite (ERTS- 1) — was launched. The purpose of ERTS-I was to gather information about agricultural and environmental conditions. This satellite was renamed Landsat 1 and was the first of three satellites that made up the first Landsat generation (Landsat 1-3).

Landsat 1 carried a return-beam vidicon with an 80m resolution and an 80 m MS sensor. In 1982, a second generation Landsat series was started with Landsat 4 followed by Landsat 5 in 1984. These satellites each had 30 m and 80 in MS sensors. Landsat 6 was launched in 1993 but was lost when its launch vehicle failed. The most recent member of the series, Landsat 7, was launched in 1999.

The first three Landsat satellites were built and launched by NASA, and they were operated by NASA until 1979. At that point, management was transferred to the National Oceanic and Atmospheric Administration (NOAA) of the Department of Commerce with the hope that a commercial remote sensing industry would develop.[13] NOAA also assumed responsibility for developing Landsat 4 and 5. The early Landsat program did much to show the value of remote sensing and fostered the development of private sector firms dedicated to extracting useful information from the Landsat images for commercial purposes.[14] In 1984, Congress passed the Land Remote-Sensing Commercialization Act of 1984 (P.L.98-965). The Act enabled the establishment of Earth Observing Satellite Corporation (EQSAT), which assumed operational responsibility for the existing Landsat satellites and, with federal assistance through NOAA, was to develop and operate Landsat 6 and 7. A dispute ensued about the funding of the new Landsat systems, and only Landsat 6 was built under the NOAA/EOSAT arrangement. Much of the dispute revolved around whether public or private use of the Landsat data should have primacy.[15] There were also concerns about whether federal contributions to the Landsat program would be adequate to ensure its continuity and whether there was adequate demand to support the satellites in the private market.

Pressure built up in the early 1990s to return Landsat operation to the federal government. In 1992, the Land Remote Sensing Policy Act of 1992 was passed which repealed the 1984 Act and transferred responsibility for the Landsat system to NASA and the Department of Defense (DOD) from NOAA and EOSAT.[16] One of the important goals of this

---

[12] The CORONA project ended in 1972. The images were declassified in 1995 and made available to the public. Gibson, *Introductory Remote Sensing,* 6-7. For a detailed history of Corona, see Dwayne A Day, et.al., *Eye in the Sky. The Story of the Corona spy Satellites,* Smithsonian Press (Washington) 1998.

[13] For further discussion of the early history of Landsat privatization, see Congressional Research Service, *The future of land remote sensing satellite system (Landsat),* by David Radzanowski, *91-685,* September 16, 1991.

[14] Ray Williamson, "Remote Sensing Policy and the Development of Commercial Remote Sensing." *Commercial Observation Satellites,* 40.

[15] Ibid., 43.

[16] As noted above, this action did not change federal policy in support of a commercial remote sensing industry; rather it recognized the possibility that such an industry would take longer to develop than was believed in the early 1980s.

legislation was to ensure continuity of Landsat data. Soon after this legislation was passed, DOD withdrew from the arrangement.[17] Landsat 7, which was developed and launched by NASA as part of its Earth Observing System (EOS) program, is now operated jointly by NASA and USGS. USGS also operates Landsat 5, the only other operating Landsat satellite.

In addition to the Landsat series, other civilian remote sensing satellites have been launched since the early 1980s. The most well known of these are the three SPOT satellites — SPOT 1, SPOT 2, and SPOT 4 — launched in 1986, 1990, and 1998 respectively.[18] These satellites were developed and are operated by Spot Image, a French company. Each of these satellites have 10m PAN and 20m MS sensors. A fifth satellite, SPOT 5 with 2.5m PAN and 10m MS sensors, is planned for launch in 2002. SPOT was the first commercial remote sensing satellite and helped fuel optimism about the development of a strong commercial remote sensing market in the early 1990s.[19]

Most recently, Ikonos 2, developed by Space Imaging, was launched. As noted, it supplied images of the September 11 attack on the World Trade Center and currently supplies images of the war in Afghanistan.[20] Two other U.S. companies Earth Watch and Orbimage have plans to launch high resolution satellites with both PAN and MS (or hyperspectral) sensors.[21] Both, however, have attempted to launch such satellites in recent years and have failed.

In addition to these satellites, other countries have or are planning to launch high resolution remote sensing satellites. In addition to India, a joint U.S./Israeli venture successfully launched a 1 meter (PAN) satellite, EROS 1A, in December 2000 and several other Countries have recently launched or are planning to launch over the next few years high resolution remote sensing satellites.[22]

ISSUES

Commercial Market Potential and Government Role

As noted above, development of a healthy commercial satellite remote sensing industry has, been a long-term goal of the federal government. While attempts to create such a market by the commercialization of Landsat in the 1980s were not successful, several factors led to a new attempt in the 1 990s. These factors included the emergence of lower-cost access to space, the use of commercial remote sensing data during the Gulf War, expansion of satellite remote sensing by foreign governments, growth in sales from SPOT, and concern about the health of the U.S. aerospace industry following the end of the cold war.[23] This interest culminated in Presidential Decision Directive 23 (PDD-23), issued on March 24, 1994, which set forth guidelines for anyone applying for a remote sensing satellite operating license under

---

[17] For further discussion of this period in remote sensing policy, see Congressional Research Service, *Public and Commercial Land Remote Sensing From Space. Landsat 7, Lewis and Clark, and Private Systems,* by David Radzanowski, 95-346 SPR, February 23, 1995.

[18] A third satellite, SPOT 3, was launched in 1993 but ceased operation in 1996.

[19] Kevin O'Connell and Beth E. Lachman, *Commercial Observation Satellites,* 69,

[20] Ikonos images can be seen on httjp://www.spaceimaging.com/level1/index33.htm

[21] Frank Sietzen, "Advanced imagery," 39.

[22] See Frank Sietzen, Jr., *Aerospace America,* 39, for a list.

the 1992 remote sensing act. These guidelines cover any activity that involves foreign access to images, technology, and systems.[24] The intent of the directive was to clear up uncertainties about license conditions that had existed since the 1992 Act. One feature of the directive was that the federal government would have the right to limit, for national security reasons, a licensee's satellite observations. This "shutter control" provision has been controversial and will be discussed more fully below.

With the issuance of PDD-23, the number of license applications grew significantly. At least 11 licenses have been granted by the U.S. Department of Commerce since 1994, including those for three satellites that are now operating — Ikonos 2, Microlab 1 (Orbview-1), and SeaStar (Orbview-2)[25]. The latter two satellites were developed by Orbimage, a division of Orbital Sciences Corporation (OSC). Three other satellites granted licenses were launched but failed to achieve operational status. They were Early Bird, launched in 1997, and Quick Bird 1, launched in 2000. Both were developed by Earth Watch. In the first case, the satellite achieved orbit but lost power five days after launch. In the second case, the satellite failed to reach orbit. In September 2001, Orbview-4, developed by Orbimage, was launched but failed to reach orbit. This satellite had a 1 meter (PAN), a 4 meter (MS), and an 8 meter hyperspectral (HS) sensor suite.[26]

In addition to regulating domestic commercial remote sensing satellites, the U.S. government is the largest customer of images from those satellites. Acting at the direction of Congress, the major users of these images, NASA and the National Imagery and Mapping Agency (NIMA) of the Department of Defense (DOD), established programs to support the development of a commercial remote sensing industry.[27] NASA purchases commercial imagery to supplement its own satellite remote sensing program for its Earth Science Enterprise. Responsibility for these purchases lies with the Commercial Remote Sensing Project (CRSP) created by NASA to accelerate "the development of the U.S. remote sensing industry."[28] In addition to purchasing images for NASA's scientific activity, CRSP is establishing cooperative agreements with private companies to help expand the use of commercial remote sensing GIS applications. NIMA has created a Commercial Imagery Program to help industry develop capabilities to provide it with products to assist its mapping mission. In addition, NIMA has established contracts with several remote sensing firms for the purchase of image products and services. In both cases, actions are being taken by the agencies to support the development of a commercial industry that can be relied upon as a steady supplier of imagery.

Despite the optimism and federal activities described above, the outlook for commercial remote sensing is uncertain. Currently annual worldwide revenues are estimated to be from $300 million to $2 billion depending on what is included. The higher figure includes all

---

[23] Kevin O'Connell and Beth E. Lachman, *Commercial Observation Satellites*, 56.

[24] See http://www.fas.org/irp/offdocs/pdd23-2.htm.

[25] Microlab 1 is operated by Orbital Sciences Corporation (OSC), but carries instruments for experiments funded by NASA and NSF. SeaStar is also operated by OSC and carries a NASA instrument, SEAWIFS, designed to monitor ocean conditions. OSC retains the rights to the data, however, and is marketing information on ocean fishing locations.

[26] As a result of the failure, Orbimage filed for Chapter 11 bankruptcy. "Orb image announces financial restructuring following launch failure," Aerospace Daily, Sept. 26, 2001, 3.

[27] P.L. 105-3-3; Sec. 107 for provisions about NASA.

[28] Kevin O'Connell and Greg Hilgenberg, "U.S. Remote Sensing Programs and Policies," *Commercial Observation Satellites*, 158.

analytical and consultant services provided in addition to the raw images. At present, no commercial remote sensing satellite owner is profitable.[29]

There appear to be several reasons for the current uncertain state of the industry. Image prices are relatively high which has restrained demand, particularly from smaller users such as academic researchers, local governments, and small businesses that might have a need for high resolution images. Evidence to date suggests a strong price sensitivity in the market.[30] Second, aerial remote sensing continues to present stiff competition, particularly as it adopts new technologies that enable lower cost, higher resolution digital images. Currently, aerial remote sensing makes up about 90% of the commercial remote sensing market.[31] A third factor is the level of value-added services that facilitate the application of remote sensing images. Given the limited expertise in interpreting and assessing remote sensing images by many potential uses, such value-added services will be crucial to the future of the remote sensing industry. A similar situation existed with the development of geographic information systems (GIS). As GIS technologies matured, processes were developed that made them easier to use and, consequently, resulted in an expanded market.[32] The natural connection between remote sensing imagery and GIS may be an important stimulus in developing remote sensing markets.

A fourth factor is the role of the federal government as a purchaser of images. In particular, for various reasons, the NIMA strategy described above does not appear to be providing the stimulus hoped for. In a recent review of that strategy, NIMA officials reaffirmed the agency's desire for a "robust space-based [remote sensing] capability", but also stated that it does not have the funds to buy as much commercial imagery as it desires.[33] This lack of funds has hurt the industry, which was counting on strong support from the federal government in the form of purchases of products and services. Without such support, industry officials claim it will be hard to raise the capital needed to build and launch the next generation of commercial remote sensing satellites, which includes a 0.5 m resolution (PAN) system planned for 2004.[34]

Others argue that NIMA does not have adequate resources to process and disseminate the images it already receives and will receive from the National Reconnaissance Office (NRO) and other sources in the next several years.[35] As a consequence, these analysts suggest that it is not reasonable to expect that more funds should be added to the NIMA budget to buy more images.[36] At the same time, these observers note that the commercial remote sensing industry has not expanded its commercial market enough to be able to survive without substantial purchases from NIMA. Others, while acknowledging NIMA's situation, suggest that NIMA should look to the commercial remote sensing industry for the technologies, products, and

---

[29] Barnaby J. Feder, *New York Times*, F1.

[30] Ibid., F7.

[31] Nick Jonson, "Make or break year for satellite imagery providers, analyst says," *Aerospace Daily*, June 7, 2001, 4.

[32] Kevin O'Connell and Beth E. Lachman, *Commercial Observation Satellites*, 66.

[33] Catherine MacRae, "Updated commercial imagery strategy cites need to back industry," *Inside the Pentagon*, July 26, 2001, 1. See also, Congressional Research Service. *US. Space Programs: Civilian, Military, and Commercial*, by Marcia Smith, IB9201 1, 8.

[34] Ibid, 2.

[35] The NRO is a DOD department that builds and operates U.S. reconnaissance satellites. It is the source of most of the data NIMA disseminates.

[36] Beth Larson and Kirk McConnell, "The Problem with Commercial Imagery," *Space News*, July 23, 2001, 13.

services that can assist with processing and dissemination rather than more images.[37] They note that rapid advances are being made in spatial image-related technologies primarily in commercial market, and that NIMA should take advantage of these changes. Furthermore, they argue that it is not certain that the commercial remote sensing industry needs substantial federal investment to survive, but rather the U.S. government should "offer a clear commitment and stick with it ... ."[38]

While NASA, acting at the direction of Congress (P.L. 105-303), also has adopted a strategy of supporting commercial remote sensing, its support of the industry is likely to be significantly smaller than NIMA's. The requirements of NASA's Earth Science Enterprise (ESE), which is responsible for the agency's remote sensing activities, are typically very specialized and require sensors that are not usually found on commercial satellites. For example, NASA is interested in measuring climate parameters, such as atmospheric water vapor, temperatures, and wind speeds, for its study of global change. In the cases where NASA is using commercial remote sensing data, the satellites are carrying specially designed NASA instruments.

Other federal agencies also use remote sensing data. In a report recently released by the Senate Governmental Affairs Committee, Subcommittee on International Security, Proliferation, and Federal Services, it was reported that of 19 civilian agencies polled, 15 used remote sensing data in some form.[39] The agencies also expressed concern about their ability to use such data as effectively as they would like. In particular, they cited high costs of commercial data, processing, and analysis; and lack of technology and trained personnel to fully exploit imagery.

In addition to purchasing commercial remote sensing images, the federal government also markets images and may act as a competitor to the commercial industry. Landsat 7, operated by NASA and USGS, makes available its images to the public at cost. Some claim that the availability of Landsat 7 images at cost has helped expand the potential market for remote sensing data.[40] The availability of raw images from Landsat also appears to help the commercial remote sensing industry by forcing users of Landsat images to go to commercial value-added firms for processing and analysis. Recently, however, USGS has proposed offering levels of enhancement as well as the raw data.[41] Many value-added firms argue that such actions would be harmful to their segment of the industry,[42] and that the 1992 Land Remote Sensing Act restricts the sale of Landsat images to "unenhanced" data.

A debate is developing over who should build and operate the next Landsat mission. In order to ensure continuity of Landsat data, NASA has stated that a new satellite — the Landsat Continuity Mission — will likely be needed by the end of the decade when the design life of Landsat 7 is complete. Currently, NASA has stated that it wants that follow-on mission to be financed, built, and operated by a private company.[43] In addition, the 1992 Land Remote Sensing Act, the 1998 Commercial Space Act, and the 2000 NASA Authorization

---

[37] Kevin M. O'Connell, "The Problem Without Commercial Imagery," *Space News*, September 3, 2001, 15.
[38] Ibid.
[39] Report by the Senate Committee on Governmental Affairs, Subcommittee on International Security, Proliferation, and Federal Services, *Assessment of Remote Sensing Data Use by Civilian Federal Agencies*, December 12, 2001.
[40] Ray A. Williamson, Commercial Observation Satellites, 49.
[41] Jason Bates, "Satellite Imagery Firms Rally Against Landsat 7 Proposal," *Space News*, September 10, 2001, 4.
[42] Ibid.
[43] Warren Ferster, "Following Landsat 7, NASA Wants Commercial Imager," *Space News*, April 12, 1999.

Act express preference for a commercial Landsat Data Continuity Mission. Several remote sensing advocates in the private sector also support a private Landsat 7 follow-on.[44] These preferences are based on the belief that a privately owned Landsat will best promote the policy of developing a commercial remote sensing industry, and that imagery from such a system will cost less to the federal government. Others, however, argue that this Landsat follow-on mission should be built by NASA. They point out that at a workshop held by NASA on the Continuity Mission, it was the consensus of commercial remote sensing image providers at the meeting that there was insufficient promise of an adequate financial return from a Landsat 7 follow-on to justify private investment.[45] In addition, members of the value-added segment of the industry in attendance argued that a privately owned Landsat that could set prices for images as it chose could prove detrimental.[46] They noted that the availability of Landsat 7 data at cost has been very helpful for in stimulating growth of the value-added segment.

The U.S. government is also influencing the development of the commercial remote sensing industry through its regulation. In particular, the possibility that the federal government may impose restrictions on what a licensee can observe and transmit, has added a measure of uncertainty to the market. While shutter control has not been invoked to date, the possibility may create concern on the part of potential commercial purchasers of remote sensing imagery that they may lose access to those images during times of national emergency.

A fifth factor is the effect of foreign competition. In addition to the French, Indian, and Israeli launches mentioned above, several other foreign remote sensing satellites are either operating or are planned to be launched within the next several years.[47] These satellites, which are government owned or are highly subsidized by the government, are or will be offering imagery and services similar to those of U.S. commercial remote sensing satellites. Competition from these entities is likely to be substantial and could seriously hinder the growth of a domestic commercial industry.

It is clear that while the market for commercial remote sensing is growing, there is little expectation that the industry will be profitable anytime soon. Nongovernmental applications have been slower to develop than anticipated, and competition from aerial remote sensing and foreign competitors is substantial. It is possible that the new generation of 1 meter (PAN) or better satellites may provide the stimulus needed to develop new, profitable applications. Until that happens, however, the economic future of the industry is uncertain.

National Security and Commercial Satellites

Commercial remote sensing satellites offer both benefits and concerns for national security. The U.S. government has made and is making extensive use of these satellites in a variety of national security-related activities. In addition to those noted above, U.S. intelligence officials are purchasing commercial images to supplement those obtained from their own satellites. Although the latter have much higher resolution, they are spread thin and

---

[44] Dr. Murray Felsher, "Space Imaging's Good Work," *Space News*, July 23, 2001, 13.
[45] Joanne Irene Gabrynowicz, "Law, Practicality and Landsat Data Continuity," *Space News*, February 19, 2001, 13.
[46] Ibid.
[47] See Frank Sietzen, Jr., *Aerospace America*, 39, for a list.

the highest resolution is not always necessary.[48] Furthermore, costs of processing commercial images are generally less than obtaining images from intelligence satellites, and the commercial satellites are often in a better position because of greater coverage. In another action, NIMA, immediately after the start of bombing in Afghanistan, contracted with Space Imaging for exclusive use of all images taken by Ikonos 2 of that country for the duration of the conflict.[49] In a related move, DOD requested $1 billion to purchase commercial satellite imagery from the $40 billion emergency funding package approved by Congress following the September 11 terrorist attacks.[50]

Nevertheless, risks remain about the growing presence of high resolution, commercial remote sensing satellites. Indeed, the existence of these risks appears to be a major reason for NIMA's action in contracting for exclusive use of the Ikonos 2 images from Afghanistan. Potential adversaries could use remote sensing imagery to determine U.S. military preparations and operations.[51] They could also use such imagery to facilitate military actions against another state or against insurgent groups within their borders. While terrorist organizations could in principle use remote sensing imagery, it may not provide much added value to information it can obtain in other ways. A more critical concern is the possibility that commercial remote sensing could be used by terrorists to compromise U.S. counter terrorism moves.[52] There is also the risk that access to commercial remote sensing imagery may permit adversaries or related groups to devise ways to limit the effectiveness of U.S. efforts to use remote sensing to monitor their activities.

In the near term, these risks do not appear to be as serious as they are likely to be in the longer term.[53] At present, most countries in currently sensitive areas do not have the capabilities for processing and analyzing remote sensing images required to make full use of their potential. Other near-term factors that may limit national security risks are that short turn around access less than 24 hours — is not likely to be available to non-satellite nations for several years; extensive cloud cover limits the value of conventional remote sensing in many areas; and processing technology for making full use of remote sensing imagery may be many years off.

A recent announcement by the Gulf Cooperation Council, a group of Arab states in the Persian Gulf region, that they are exploring the possibility of purchasing their own remote sensing satellite suggests, however, that some of these mitigating factors may be eliminated or greatly reduced relatively soon.[54] In addition, technological advances such as the penetration of high resolution radar satellites and more automated image processing will help overcome some of the other factors.

The expansion of foreign remote sensing systems could present the United States with significant national security risks. Should that happen, it maybe necessary have to consider options to neutralize as much of the risk as possible. Such actions, however, would have to be weighed against harm that could come about by reducing the potential national security

---

[48] Barnaby J. Feder, *New York Times,* F7.

[49] Pamela Hess, "DOD locks up commercial satellite pix," *United Press International,* October 12, 2001.

[50] Tony Capaecio, "U.S. Pentagon Asks $19 Bln For Weapons, Intelligence," *Bloomberg. com,* September 19, 2001.

[51] John C. Baker and Dana J. Johnson, "Security Implications of Commercial Satellite Imagery," *Commercial Observation Satellites,* 116.

[52] Ibid., 117.

[53] Ibid., 120-125.

[54] Warren Ferster and Gopal Ratnam, "Gulf States Consider Buying Spy Satellite," *Space News, December 10, 2001, 1.*

benefits of increased accessibility to high resolution remote sensing satellites. In particular, as many analysts have suggested, such systems could make the planet more transparent, leading to a world where malevolent intentions are more easily observed and thwarted, and, thus they could possibly act as a significant deterrent to conflict.[55]

An indirect national security concern could arise as a result of misuse of remote sensing imagery by individuals and organizations that are not adequately trained to interpret those images. Interpreting remote sensing images is difficult and involves detection, identification, measurement, and analysis. Without proper equipment or training, incorrect interpretation of images can result, particularly if they are somewhat ambiguous in the first place.[56] Differences of opinion even contentious debate could develop over competing interpretations of their meaning.

Such debates can have implications for national security because many times they are about images of objects of military significance that are the basis of proposed foreign policy or other actions. Examples include satellite images of long-range missile development in North Korea and nuclear weapons testing in Pakistan and India.[57] If the debates are the result of misinterpretation by untrained analysts, they could create credibility problems with the imaging process and undermine U.S. foreign policy. These concerns could be mitigated by cooperative arrangements between experienced and newer, untrained users of remote sensing images. This cooperation could be particularly beneficial for the news media and non-governmental organizations.

Regulation and Commercial Remote Sensing: Shutter Control

The Land Remote Sensing Act of 1992 is the basis for regulation by the federal government of the commercial remote sensing industry.[58] According to the 1992 Act, a license must be granted before an entity can begin operating a remote sensing satellite. The licensee must protect U.S. national security, observe international obligations of the United States, make raw images available at a reasonable cost to the governments of the lands observed, and notify the Secretary of Commerce of any significant international agreements entered into by the licensee. The Act also prohibits use of remote sensing to gather, transmit, or deliver defense related information for the benefit of any foreign government or to publish photographs of defense facilities. As noted above, Presidential Decision Directive 23 (PDD-23) was issued in 1994 to specify the policy of the last administration towards commercial remote sensing. The directive allows the Secretary of Commerce to limit image collection and distribution by licensed remote sensing satellites for national security reasons, the so-called shutter control provision. The policy outlined in PDD-23 also promotes the remote sensing industry and provides for low cost access to the images for the United States.

There remains ambiguity and uncertainty in the remote sensing industry about the current regulatory framework governing commercial remote sensing in this country. Rules

---

[55] John C. Baker, Kevin M. O'Connell, and Ray A. Williamson, *Commercial Observation Satellites*, 563.
[56] John C. Baker, "New Users and Established Experts: Bridging the Knowledge Gap in Interpreting Commercial satellite Imagery," *Commercial Observation Satellites,* 542.
[57] Ibid., 534, 543.
[58] The discussion in this section is based primarily on Bob Preston, "Space Remote Sensing Regulatory Landscape," *Commercial Observation Satellites,* 501.

implementing the Act have taken over eight years from its enactment to develop, with an interim final rule published in July 2000. An attempt was made to clarify the regulations during the 105th Congress, but no amendments to the 1992 Act were included in the Commercial Space Act of 1998 (P.L.105-303). The 1998 Act did, however, reiterate Congress's intent that the Federal government pursue opportunities to purchase commercial remote sensing imagery when possible.

At issue in the regulations is whether the doctrine of prior restraint can be used to limit actions by remote sensing entities — shutter control and whether restrictions can be imposed based on foreign policy considerations. In the first case, the debate centers on whether there is adequate justification for regulations that prevent the publication of images prior to any event that may justify that action. Some argue that concerns that national security could be endangered by publication of such images may not be sufficient to justify the use of prior restraint. They argue that it is necessary to prove that application of shutter control would prevent the loss of life before a prior restraint regulation could be established. Others argue that national security considerations are sufficient, and that the use of prior restraint is justified even if the level of danger cannot be specified in advance.

There is also debate about whether the Act authorizes the use of foreign policy concerns as a justification for shutter control prohibiting the acquisition and publishing of remote sensing images of sensitive areas. PDD-23 does include foreign policy along with national security as a justification for shutter control. Some have argued that the term foreign policy in this context is too vague and a more precise expression is needed. These observers suggest the basis for shutter control should be protection of international obligations when harm is possible if images are obtained and published.

The crux of this issue is how to balance the seemingly conflicting goals of promoting a commercial remote sensing industry while also protecting national security. The difficult regulatory challenge is to prevent inappropriate use of remote sensing in gathering information on national security sites for a possible adversary without excessively hindering legitimate commercial operation. While prior restraint through the application of shutter control would likely achieve the first goal, its application could cause severe financial harm for the entity targeted. As noted above, however, the U.S. government has taken another path in the current situation in Afghanistan by contracting with Space Imaging for exclusive purchase of the Ikonos 2 images. This has the advantage of preserving the commercial livelihood of the firm while denying access to potentially critical images by adversaries of the United States. It also has the advantage of avoiding a likely court challenge over the legality of using prior restraint in establishing shutter control. The action has not been spaced criticism, however, as some have claimed the move is more to control information flow to the public than assure access to strategic images.[59]

Whether such "commercial" actions can be taken in all cases where shutter control might be invoked seems problematic. If not, there does not appear to be any other option than prior restraint. Even then, a court challenge is likely. Furthermore, as the number of high-resolution remote sensing satellites outside the control of the Unites States grows, the application of shutter control or exclusive purchase contracts will become less and less effective in preventing unwanted access to sensitive images.

---

[59] Pamela Hess, *United Press International*, October 12, 2001.

POLICY AND CONCLUSIONS

To date, federal policy directed at facilitating the development of a strong commercial remote sensing industry appears to have had mixed success. While an industry exists, its future appears uncertain. There are at least three options that might improve the outlook should that remain a national policy goal. They are federal investment, applications development, and regulatory stability.

It seems clear that federal investment in the industry at some level is likely to be required for an indefinite period. While the primary source of this support is now NASA, NIMA, and the intelligence community, Congress may wish to encourage expansion of that base to other agencies such as the Departments of Agriculture, Interior, and Transportation. As noted in the discussion about NIMA's role, such support may be most beneficial if it includes processing and interpretive services as well as the collection of imagery. In that way, the important value-added sector of the industry would also receive assistance. Some observers have argued that an important condition for such investment is that it focus on the missions of the funding agencies and not be set up primarily to develop the industry.[60]

In a related matter, Congress may wish to reexamine the Landsat Data Continuity Mission. While there appear to be strong arguments supporting private sector development, depending on the state of the industry when the satellite is needed, private development may not be practical. It may be necessary for NASA to provide some or all funding for the project if it is to be built in a timely manner. Arguments by segments of the remote sensing industry suggest such a situation would still be beneficial. Whatever decision is made, it appears important that the system maximize support for the industry while maintaining broad public access to images.

Congress may also wish to expand its consideration of applications development. Perhaps even more important than federal investment is the development of remote sensing applications that would drive an expanded market. The value-added and applications research component of the U.S. commercial remote sensing industry is an important reason for the U.S. lead in this industry. To maintain that lead, it will be important that industry is encouraged to continue to seek innovative applications. Applications that would permit small users that do not possess the technology or resources of large users such as NIMA to make full use of the potential offered by these images maybe of particular importance. Congress may wish to provide funds to relevant agencies to support research and development for new remote sensing applications that could be of value to state and local governments and to non-governmental organizations.[61] The Remote Sensing Applications Act of 2001, H.R.2426 sponsored by Rep. Udall and introduced on June 28, 2001, would direct NASA to: "establish a program of grants for pilot projects to explore the integrated use of sources of remote sensing and other geospatial information to address State, local, regional, and tribal agency needs ..."

In a recent report, the National Research Council (NRC) of the National Academies found that there were several "gaps that must be bridged" if effective civilian applications of

---

[60] Kevin M. O'Connell, *Space News.*
[61] Space Enterprise council, "Promote a Competitive U.S. Commercial Remote Sensing Industry," *U.S. Chamber of Commerce,* (Washington, DC) July 2001.

remote sensing were to be developed.[62] Included were the gaps between the raw remote sensing data and useful information, between the expertise of experienced suppliers and users of remote sensing data and potential new users, and between the acquisition cost of raw remote sensing data and the cost of products that permit full utilization of that data. To address these gaps and enhance applications development, the NRC made several recommendations. Among them are that NASA's Earth Science Enterprise develop cost-benefit analyses of a range of remote sensing applications and the costs of implementing those applications; that NASA, NOAA, the Department of Agriculture, and USGS provide funds to develop training materials and courses for new users of remote sensing data; that NASA, NOAA, USGS, and other relevant civilian agencies support remote sensing applications research; that data producers establish mechanisms to obtain feedback and advice from the user community; and that data standards and protocols be established.

Much of the responsibility for developing such applications for the commercial sector, however, will likely fall on the industry itself. One observer has suggested that partnerships between companies that operate satellites and those that provide the value-added services maybe helpful in this connection.[63] Development of applications and processing that reduce the cost of using imagery and take advantage of the unique contribution of satellite remote sensing will be particularly important.

Congress may also wish to consider ways to improve the regulatory environment. A stable regulatory environment appears to be important for future growth of the commercial remote sensing industry. While concern has been expressed about the current ambiguity in regulatory procedures, that uncertainty has not prevented the several licensees that have been granted since 1994. Nevertheless, uncertainty is not conducive to long term growth of the industry. The key factor is the application of shutter control. If shutter control is invoked, it seems important that it be done in a way that minimizes disruption by offering alternatives if possible and that clearly spells out the reasons *for* the constraints. Some have also suggested that foreign policy not be used as a reason for invoking shutter control, but rather specific reference should be made to protection of "U.S. diplomatic forces, installations, and operations."[64] While this change would provide a more specific basis for applying shutter control, it may leave out important national considerations that involve a broader definition of foreign policy such as meeting international obligations.

Regulatory stability is also likely to be important when considering emerging technologies. In particular, as hyperspectral and synthetic aperture radar (SAR) sensors gain in resolution, they may pose new regulatory challenges because of their capabilities with respect to national security considerations. Currently, the United States limits commercial SAR resolution to 5 meters. Foreign SAR systems such as Canada's proposed RADARSAT 2, however, are planned with higher resolution and maintaining the 5 meter limit may hinder U.S. competitiveness in this area. A regulatory scheme that recognizes the realities of the international market may be necessary.

The commercial remote sensing industry has not developed as rapidly as proponents have. Furthermore, the U.S. lead in commercial remote sensing may be threatened by growing foreign competition. Nevertheless, an international commercial remote sensing

---

[62] The National Research Council, The National Academies, *Transforming Remote Sensing Data into Information and Applications.* (Washington, DC) December 2001.
[63] Nick Jonson, *Aerospace Daily.*
[64] Bob Preston, *Commercial Observation Satellites,* 521.

industry is a reality and is changing the way many government and non-government entities approach geospatial and intelligence matters. As pointed out by the authors of *Commercial Observation Satellites,* this presence appears to be ushering in an age of "global transparency." As sensor resolution and the technology for processing and interpreting data improves, this transparency will grow. Any Congressional actions to address the future of U.S. commercial remote sensing likely will have to address the ramifications of these international transparency and competition realities.

Chapter 10

V-CHIP AND TV RATINGS: HELPING PARENTS SUPERVISE THEIR CHILDREN'S TELEVISION VIEWING

Marcia S. Smith

SUMMARY

To assist parents in supervising the television viewing habits of their children, Congress included a provision in the Telecommunications Act of 1996 (P.L. 104-104) that new television sets with screens 13 inches or larger sold in the United States be equipped with a "V-chip" to screen out objectionable programming. As of January 1, 2000, all such TV sets must have a V-chip. Use of the V-chip by parents is optional. In March 1998, the Federal Communications Commission approved a ratings system that had been developed by the television industry to rate each program's content, which enables the V-chip to work. Congress and the FCC have been monitoring implementation of the V-chip. Some are concerned that it is not effective in curbing the amount of TV violence viewed by children and want further legislation. S. 341 (Hollings) and H.R. 1005 (Shows) would require a study of the V-chip's effectiveness, and, if it is not effective in curbing children's viewing of TV violence, a "safe harbor" time period when violent programming could not be telecast would be created. This report will be updated if warranted.

REQUIREMENT FOR A V-CHIP

Section 551 of the Telecommunications Act of 1996 (P.L. 104-1 04, February 8, 1996) requires that all new television sets with a picture screen 13 inches or greater (measured diagonally) sold in the United States be equipped with a device that can block certain television programming. Dubbed the *"V-chip"* for "violence chip," the intent is to give parents more control over what their children see on television. On March 12, 1998, the Federal Communications Commission (FCC) Set January 1, 2000 as the date by which V-chips must be installed in such TV sets. The FCC also adopted technical standards for the V-chip and approved the industry-developed ratings system (see below) that enables the V-chip to work Some companies plan to offer devices that can work with existing TV sets.

The V-chip is a computer chip that reads an electronic code transmitted with the television signal (cable or broadcast) indicating how a program is rated. Using a remote control, parents can enter a password and then-program into the television set, which ratings are acceptable and which are unacceptable. The chip automatically blocks the display of any programs deemed unacceptable. The ratings data are sent on line 21 of the Vertical Blanking Interval found in the National Television System Committee (NTSC) signals used for U.S. television broadcasting. Use of the V-chip by parents is entirely optional.

ESTABLISHING A RATINGS SYSTEM

The first step in implementing the law was creating a ratings system for television programs, somewhat similar to how movies have been rated since 1968 by the Motion Picture Association of America (MPAA). The law urged the television industry to develop a voluntary ratings system acceptable to the FCC, and the rules for transmitting the rating, within one year of enactment. Although the "V" is for violence, the ratings system actually is intended to reflect "sexual, violent or other indecent material about which parents should be informed before it is displayed to children, provided that nothing in this paragraph should be construed to authorize any rating of video programming on the basis of its political or religious content" [section 551(b)(1)].

After initial opposition, media and entertainment industry executives met with President Clinton on February 29, 1996, and agreed to develop the ratings system because of political pressure to do so. Many in the television industry have been opposed to the V-chip, fearing that it will reduce viewership and hurt advertising. They also question whether it violates the First Amendment. Industry executives said they would not challenge the law immediately, but left the option open for the future (the law provides for expedited judicial review).

Beginning in March 1996, a group of television industry executives[1] under the leadership of Jack Valenti. President of the MPAA (who created the movie ratings), met to develop a TV ratings system. On December 19, 1996, the group proposed six age-based ratings (TV-Y, TV-Y7, TV-G, TV-PG, TV-l4 and TV-M) including text explanations of what each represented in terms of program content. In January 1997, the ratings began appearing hi the upper left-hand corner of TV screens for 15 seconds at the beginning of programs, and were published in some television guides. Thus, the ratings system was used even before V-chips were installed in new TV sets. News shows and sports are not rated (the Valenti group does not consider talk shows or programs about show business and reports on public figures and other issues of general interest to be news). All other programs are rated by the broadcast and cable networks and producers of programs. Local broadcast affiliates can override the rating given a particular show.

---

[1] The group included the national broadcast networks; independent, affiliated and public television stations; cable programmers; producers and distributors of cable programming; entertainment and movie studios; and members of the guilds representing writers, directors, producers and actors.

Critics of the December 1996 proposal, including some Members of Congress and groups such as the National Parent-Teacher Association, argued that the ratings provided no information on why a particular program received a certain rating. Some advocated an "S-V-L" system (sex, violence, language) to indicate with letters why a program got a particular rating, possibly with a numeric indicator or jointly with an age-based rating. Another alternative was the Home Box Office/Showtime system often ratings such as MV (mild violence), V (violence), and GV (graphic violence). Critics also argued that having industry rate its own programming lacked credibility. Mr. Valenti countered there is no practical alternative to rating approximately 2,000 hours of programming per day.

THE CURRENT "S-V-L-D" RATINGS SYSTEM

In response to the criticism, most of the television industry agreed to a revised ratings system (see box) on July 10, 1997, that went into effect October 1, 1997. The revised ratings system adds designators that indicate whether a program received a particular rating because of sex (S), violence (V), language (L), or suggestive dialogue (D). A designator for fantasy violence (FV) was added for children's programming in the TV-Y7 category. On March 12, 1998, the FCC approved the revised ratings system, along with V-chip technical standards, and the effective date for installing them (discussed earlier).

U.S. TELEVISION INDUSTRY'S REVISED TV RATINGS SYSTEM

The following categories apply to programs designed solely for children:

TV-Y: *All Children. This program is designed to be appropriate for all children.* Whether animated or live-action, the themes and elements in this program are specifically designed for a very young audience, including children from ages 2-6. This program is not expected to frighten younger children.

TV-Y7:   *Directed to Older Children. This program is designed for children age 7 and above.* It may be more appropriate for children who have acquired the developmental skills needed to distinguish between make-believe and reality. Themes and elements in this program may include mild fantasy violence or comedic violence, or may frighten children under the age of 7. Therefore, parents may wish to consider the suitability of this program for their very young children. Note: For those programs where fantasy violence may be more intense or more combative than in other programs in this category, such programs will be designated TV-Y7-FV.

The following categories apply to programs designed for the entire audience:

TV-G: *General Audience. Most parents would find this program suitable for all ages.* Although this rating does not signify a program designed specifically for children, most parents may let younger children watch this program unattended. It contains little or no violence, no strong language and little or no sexual dialogue or situations.

TV-PG: *Parental Guidance Suggested. This program contains material that parents may find unsuitable for younger children.* Many parents may want to watch it with their younger children. The theme itself may call for parental guidance and/or the program contains one or more of the following: moderate violence (V), some sexual situations (S), infrequent coarse language (L), or some suggestive dialogue (D).

TV-14:   *Parents Strongly Cautioned. This program contains some material that many parents would find unsuitable for children under 14 years of age.* Parents are strongly urged to exercise greater care in monitoring this program and are cautioned against letting children under the age of 14 watch unattended. This program contains one or more of the following: intense violence (V), intense sexual situations (S), strong coarse language (L), or intensely suggestive dialogue (D).

TV-MA: *Mature Audience Only. This program is specifically designed to be viewed* by *adults and therefore may be unsuitable for children under 17.* This program contains one or more of the following: graphic violence (V), explicit sexual activity (S), or crude indecent language (L).

Source: Letter to the Federal Communications Commission submitting proposed rating system revision, August 1, 1997. (Signed by the presidents of the Motion Picture Association of America. National Cable Television Association, and National Association of Broadcasters).

In May 1995 the FCC created a V-chip Task Force, chaired by Commissioner Tristani. Among other things, the task force was charged with ensuring that the blocking technology is available and that ratings are being transmitted ("encoded") with TV programs; educating parents about V-chip; and gathering information on the availability, usage, and effectiveness of the V-chip. The task force has issued several reports and surveys [www.fcc.gov/vchip]. A February 2000 task force survey found that most broadcast, cable, and premium cable networks, and syndicators, were transmitting ratings ("encoding") and those that were not either planned to do so in the near future or were exempt sports or news networks. Of the major broadcast networks, according to the survey, only NBC does not use the S-V-L-D indicators, using the original ratings system instead. One problem impacting V-chip's effectiveness is that not all parents are aware of it. Commissioner Tristani noted in April 2000 that only 39% of parents have heard of V-chip and called on the major networks to educate parents about it.

ACTION IN THE 105TH AND 106TH CONGRESSES

During the 105th Congress, the Senate Commerce Committee held a hearing on February 27, 1997. Three bills were introduced. S. 363 (Hollings) and H.R. 910 (Markey). Often called the "safe harbor" bills, would have prohibited violent programming from being shown (with some exceptions) during hours when children comprise a substantial portion of the audience unless it could be screened out by a V-chip specifically on the basis of its violence. The Hollings bill was reported from the Senate Commerce Committee (S.Rept. 105-89). S. 409 (Coats) would have required broadcast television stations to use a content-based ratings system as a condition of obtaining or renewing their licenses.

In the summer of 1997, as debate was underway about modifying the ratings system to add the S-V-L-D designators, TV and entertainment industry representatives were concerned that Congress would attempt to pass further legislation even if they agreed to the modifications. The industry sought and was given assurances by many of the principal House and Senate critics that Congress would not move on such legislation for a period of time if the modified ratings were adopted. In a July 3, 1997, "Dear Colleague" letter, Senators Lott. McCain, Hatch, and others said there should be "a substantial period of governmental forbearance during which further legislation or regulation concerning television ratings, content or schedule should be set aside." A July 7, 1997, letter to key industry leaders signed by Representatives Markey, Burton, Moran and Spratt expressed the same sentiment but more precisely set the moratorium as 3 years beginning October 1, 1997, the date that the new ratings system went into effect. The moratorium thus has expired. Some key Senators never agreed to the moratorium, including Senator Hollings, who has continued his attempts to win passage of a safe harbor bill.

In the 106th Congress, Senator Hollings introduced S. 876 on April 26. 1999 to make it unlawful to distribute to the public any violent video programming during hours when children are reasonably likely to comprise a substantial portion of the audience, thereby establishing a "safe harbor." As introduced, the bill would have required the FCC to define that time period and the term "violent video programming." Premium and pay-per-view cable programming would be exempted and the FCC could also exempt other programming such as news and sports. The FCC would take into account whether a licensee has complied with the bill when determining whether to renew the license. The bill as introduced did not mention V-chip.

On May 13, 1999, Senator Hollings offered the text of the bill as an amendment to the juvenile justice bill (S. 254), but it was rejected (60-39). On May 18,1999, a hearing was held by the Senate Commerce Committee. With the 3-year moratorium at an end and concern about TV violence undiminished, the Senate Commerce Committee reported out a revised version of S. 876 on October 26, 2000 (S.Rept. 106-509). There was no further action on the bill. It has been reintroduced in the 107th Congress and is discussed below.

107TH CONGRESS ACTION

On February 15, 2001, Senator Hollings introduced S. 341, which is the same as S. 876, as reported, from the 106th Congress. Representative Shows introduced a House companion,

H.R. 1005. S. 341/H.R. 1005 would require the FCC to define the terms "hours when children are reasonably likely to comprise a substantial portion of the audience" and "violent video programming." Guidance is provided to the FCC on the latter definition. Furthermore, the legislation would:

- Require the FCC to report to Congress on the effectiveness of the V-chip and the ratings system in protecting children from TV violence. If the FCC finds they are not sufficiently effective, it shall initiate a rulemaking to prohibit the distribution of violent video programming during hours when children are reasonably likely to comprise a substantial portion of the audience.

- Make it unlawful to distribute to the public violent TV programming that is not blockable by the V-chip specifically on the basis of its violent content during hours when children are reasonably likely to comprise a substantial portion of the audience. (This means that the programming would have to be encoded with a "V" designator, which NBC, for example, currently does not use.) The FCC *could* exempt certain programming such as news and sports, and *must* exempt premium and pay-per-view cable and satellite programming.

Separately, H.R. 1916 (Wamp) would, *inter alia,* require labels to be placed on visual media products to indicate violent content. The bill calls for a voluntary system to be developed by manufacturers and producers of interactive video game products and services, video program products, motion picture products, and sound recording products. If a system is not developed voluntarily, the Federal Trade Commission would be required to establish such a labeling system. Thereafter, sale or distribution of such products without a label would be prohibited.

OTHER COUNTRIES

Violence on television is not unique to the United States, and other countries also have debated the V-chip concept. The V-chip is often said to have been invented in Canada.[2] The Action Group on Violence on Television (AGVOT) was charged by the Canadian Radio-television and Telecommunications Commission (CRTC) with developing a nationwide ratings system. AGVOT tested one during 1996 in which each program had a four-digit rating indicating a level of 0-5 for age, and for violent, language and sexual content. For example, a program rated 3234 would indicate age level 3 (adult 16±), violence level 2 (mild), language level 3 (coarse), and sexuality level 4 (full nudity). AGVOT withdrew this experimental system in December 1996 because of technical problems and difficulty in deciding how to rate certain programs. Others in Canada complained that it was too complicated, or that the Canadian system should be compatible with the one used in the United States since so much U.S. TV programming is seen in Canada. In April 1997, AGVOT proposed a new system that is quite similar to the Valenti group's original proposal except that it adds a category for

---

[2] A history of Canada's interest in reducing violence on television and the V-chip can be found at the following World Wide Web site: [http://www.crtc.gc.ca/eng/info_sht/tvle.htm].

exempt programming. CRTC adopted the ratings on June 18,1997. The Canadian TV Ratings (CTR) are: CTR-E (exempt, including news, sports, documentaries, talk shows, other informational programming, music videos, variety); CTR-C (for children 8 and younger); CTR-8H- (for children over 8); CTR-FAM (family viewing); CTS-PA (parental guidance); CTR-14+ (for those over 14); and CTR-1 8-r (for adults). Each is accompanied by text explaining what the rating indicates, especially in terms of violence.

Canada's implementation of the V-chip proceeded more slowly than anticipated. In March 1996, CRTC had directed the broadcast industry to encode programs with ratings beginning on September 6, 1996. By January 1997, television distributors and cable companies were to ensure that foreign signals they broadcast also were encoded (70% of Canada's television programming originates in the United States). CRTC postponed these dates in October1996, however, with rating and encoding to be in effect for the fall 1997 season and the foreign signal requirement extended until September 1997. The ratings system went into effect for English-language and specialty programming on September 29, 1997. French-language and premium programming will continue to use their own established ratings systems.

Other countries also have looked at the issue. European Union culture ministers debated the issue following a call from the European Parliament to institute a V-chip requirement, but the EC Council decided instead to study the matter further. Australia announced new censorship controls following the Port Arthur massacre in April 1996 where 35 people were killed. The massacre sparked a debate about violence on television and in the entertainment industry. Among the new controls is a V-chip requirement in new television sets.

CONCLUSION

The effect of television violence on society, especially children, has been long debated. What effect the V-chip will have is controversial. Supporters claim that since television producers will want their shows to be viewed in as many households as possible, they will reduce the level of violence in the programs. Critics complain that television will become lackluster. Others assert that if one violent scene will earn a program a bad rating, then producers will feel free to have more violence in any program since it will be blocked anyway, hence increasing the overall violence level. Others argue that many older children will be able to defeat the password-protected system and change what their parents have programmed. Or they could watch a smaller than 13-inch TV set.

Virtually everyone agrees that the V-chip is no panacea. Ultimately, parents must take responsibility for their children's viewing habits. TV ratings and the V-chip are merely tools to assist them.

Chapter 11

JAPAN'S TELECOMMUNICATIONS DEREGULATION: NTT'S ACCESS FEES AND WORLDWIDE EXPANSION

Dick K. Nanto

SUMMARY

In July 2000, the United States and Japan reach a negotiated settlement on Japan's costly rates for telecommunications companies to hook into the telephone network owned by the Nippon Telegraph and Telephone Company (NTT), Japan's dominant provider of telecom services. Japan agreed that NTT would lower its rates for regional access by 50% and local access by 20% over two years. NTT also is attempting to acquire Verio, an Internet service provider in the United States.

INTRODUCTION

Nippon Telegraph and Telephone (NTT), Japan's former domestic telephone monopoly, is being privatized, but it still is majority owned (53%) by the government and charges high rates for foreign telecommunications suppliers (including U.S. Internet providers) to connect to its telephone network. The United States claims that Japan's expensive interconnection fees violate that country's commitments under the World Trade Organization[1] and disadvantage non-NTT Internet suppliers. As part of U.S-Japan deregulation talks, the U.S. Trade Representative and others have been discussing the issue with Japan. Japan had proposed that interconnection charges by NTT be reduced a total of 22% by 2004, while the United States has pressed for a cut in rates of 41% by 2002. On July 18, 2000, the two sides

---

[1] The annex on telecommunications to the WTO's General Agreement on Trade in Services relates to measures, which affect access to and use of public telecommunications services and networks. It requires that such access be accorded to another party. on reasonable and non-discriminatory terms, to permit the supply of a service included in its schedule.

agreed that Japan would lower its rates for regional access by 50% and for local access by 20% over two years.[2]

On July 5, 2000, ten U.S. senators sent a letter to the Ambassador of Japan to urge a resolution of the telecommunications trade issue before the Group of Eight (G-8) Economic Summit in Okinawa on July 21-23, 2000. In the 106th Congress, Senate Resolution 275 (Sense of the Senate Regarding Fair Access to Japanese Telecommunications Facilities and Services) was passed and engrossed in H.R. 434 (Section 709, P.L. 106-200, signed May 18, 2000).

NTT's majority owned mobile telephone unit DoCoMo also has been building an international network primarily by purchasing shares of foreign telecommunications companies. The company already is moving into Europe and is attempting to purchase a controlling share of the Englewood, Colorado based Verio, Inc. for $5.5 billion. Verio is a major provider of Internet services to corporations. The FBI has raised national security concerns about this planned acquisition.[3] Under section 721 (the Exon-Florio provision) of the Defense Production Act of 1950 (enacted in 1988), the President is authorized to suspend or prohibit any foreign acquisition, merger, or takeover of a U.S. corporation that is determined to threaten the national security of the United States.[4] The committee reviewing the case has until August 14,2000, to either have the FBI and NTT's DoCoMo reach a satisfactory compromise or to report its decision to the President who than has 15 days to take action. On May 25, 2000, seven Republican members of the House Commerce Telecommunications subcommittee sent a letter to the USTR citing NTT's failure to institute "fair cost based interconnection rates" and criticizing the proposed Verio purchase.[5] It is unclear whether the provisions of S. 2793 (Hollings), which strengthens the limits on the holding of and transfer of broadcast licenses and other telecommunications media to foreign persons or governments, would apply to the NTT-Verio case, since Internet service providers do not need a telecommunications license.

Both NTT and NTT DoCoMo are huge companies. According to a *Business Week* ranking of global corporations, as of May 31, 2000, NTT DoCoMo was valued at $247.24 billion, which made it the world's eighth largest company (up from 27th in 1999) by market value. NTT was fifteenth (down from thirteenth in 1999) at a market value of $189.16 billion. Both NTT and NTT DoCoMo are larger than Deutsche Telekom ($187.25 billion), France Telecom ($148.71 billion), AT&T ($109.10 billion), and British Telecommunications ($93.70 billion).[6] If combined, the two NTT companies would rank as the second largest company (in terms of market value) in the world behind General Electric and ahead of Intel, Cisco Systems, and Microsoft.

---

[2] U.S. Trade Representative. United States and Japan Agree on Interconnection Rates, Press Release 00-5 5, July 18, 2000. On Internet at [http://www.ustr.gov].

[3] The FBI reportedly is concerned that its ability to enforce wiretap laws would be compromised by foreign ownership of Internet service providers. Schwartz, John. FBI Intervenes in Planned Sale of Internet Service to Japanese. *Washington Post,* July 7, 2000. P. E4.

[4] The security review is conducted by the Committee on Foreign Investment in the United States (CFIUS), which consists of eleven members from various executive branch departments and is headed by the Secretary of the Treasury. Once CFIUS has received a complete notification, it begins a thorough review of the notified transaction. If an extended review or investigation is necessary, CFIUS must begin it no later than 30 days after receipt of a notice and end within 45 days. For details on the operations of CFIUS, see its Internet site at [http://www.ustreas.gov/oasia/oii.html].

[5] Pressure Mounts on Japan on Telecommunications Competition. *Inside U.S. Trade,* June, 2000. Internet edition.

[6] The Global 1000. Special Report. *Business Week,* July 10, 2000. p. 107ff.

BACKGROUND

In U.S. economic relations with Japan, the United States has pursued a multi-faceted approach that has included encouraging major structural reform in Japan to open more sectors to competition. The global Internet revolution has pushed Internet access and the provision of Internet service in Japan to the forefront of concerns by U.S. companies. This is an area in which U.S. providers are competitive and offer leading technologies.

Since the Denver G-7 summit on June 19, 1997, the United States and Japan have been negotiating over various issues under the framework of the Enhanced Initiative on Deregulation and Competition Policy. With respect to the telecommunications sector under this initiative, Japan is to undertake specific new measures to introduce more competition into its $130 billion telecommunications market. Specifically, Japan is to ensure that interconnection rates — the rates charged competitors of NTT to access the majority of Japanese customers — are set below retail rates; define measures that will assure NTT DoCoMo's (mobile/cellular service provider) interconnection rates are more fairly priced by being purely based on costs; authorize an inter-connection "clearinghouse" for new entrants in the Japanese market which will dramatically speed market entry and liberalize the use of flexible network arrangements, thus allowing businesses to build out their networks more rapidly and efficiently.[7]

Over the more than two years that the United States and Japan have been negotiating over NTT's interconnection rates, the Japanese position has been that NTT's pricing policies are justified by its costs. The U.S. side, however, has pointed out that NTT's method of calculating costs includes too many fixed costs that have already been incurred rather than using a methodology (Long Run Incremental Cost model) scheduled to be implemented by NTT by the end of 2000 that uses only future costs in calculating rates.[8] NTT's argument that it requires more costly interconnection rates has been weakened considerably by the high rate of profit reported by its East regional telephone company.[9]

The cost question also is muddied by NTT's role as a provider of universal telephone service in Japan. As shown in Figure 1, the NTT Holding Company is organized into NTT East and NTT West regional telecommunications companies, NTT Communications (long - distance and international), and NTT DoCoMo (mobile/ cellular telecommunications). Under the regional telecom companies, NTT plans to offer Internet service, e-business, and TV broadcasting as profit seeking-activities. An important question is to what extent NTT should have to provide services to customers in remote areas at the same price charged those in cities as it has traditionally done for telephone services. The amount NTT charges other companies to connect to its lines in urban areas determines, to some extent, the funds it will have available to subsidize high-cost remote connections.

Figure 1 also shows NTT's plans for future provision of profit-seeking activities in the Internet and in broadband services (TV broadcasting). This is where interconnection fees

---

[7] U.S. Trade Representative, Government of Japan. Second Joint Status Report under the U.S-Japan Enhanced Initiative on Deregulation and Competition Policy. May 3, 1999. On the Internet at [http://www.ustr.gov/releases/1999/05/index.html].

[8] U.S., Japan Fail to agree on Telecom Deregulation, Expect New Try. *Inside U.S. Trade*, March 24, 2000. Internet Edition.

[9] Landers, Peter. Japan Signals a Truce in U.S. Phone Spat – To Resolve Trade Impasse Tokyo Seeks to speed Cut in Charges by NTT. *Wall Street Journal*, July 3, 2000. P. A9.

would become critical for foreign companies hooking into NTT's transmission system. The company currently offers ISDN (Integrated Services Digital Network) at 128 kilobits per second, a relatively slow Internet connection. NTT also offers ADSL (Asymmetric Digital Subscriber Lines) at a few megabits per second, a fast Internet connection that can handle video on demand. By the year 2003, NTT plans to offer VDSL (Very-high-bit-rate Digital Subscriber Lines at about 52 megabits per second) technology. VDSL allows a super-fast Internet connection, video on demand, interactive TV, and TV broadcasting. By 2006, NIT plans to offer FTTH (Fiber To The Home) technology with transmission speeds of 1 gigabit per second. It would carry services provided under VDSL, but at a much faster rate, as well as high-quality visual television broadcasting, medical care and educational programming delivered to an individual subscriber and other services of a future multimedia-capable home.[10] The market for these telecommunications services in Japan is large and growing. The U.S. contention is that foreign firms should have access to this market on an equitable basis and in accord with Japan's obligations under the World Trade Organization.

As part of the deregulation of Japanese companies, government-mandated divisions and government-sanctioned monopolies have been disappearing. In the case of NTT, it had held a domestic monopoly on the provision of telephone services, but it also had been barred from doing international business (done by KDD, Kokusai Denshin Denwa). Among the major carriers in the world, it is the only one that has not developed international operations. NTT now is attempting to become a global power — hence its recent deals to acquire foreign telecommunications companies.

Table 1 lists foreign acquisitions and the establishment of overseas subsidiaries by NTT. As is apparent, NTT is attempting to establish a world-wide network through focusing on Asia and on the more industrialized economies of the world. It is in the U.S. interest to ensure that U.S. and other companies have comparable access to Japan's home telecommunications market as NTT has in U.S. and other telecommunications markets in the world. Monitoring Japan's compliance with the July 2000 agreement, pressure on Japan to further deregulate its telecommunications sector, and if necessary for the United States to take unsettled issues before the dispute resolution mechanism of the World Trade Organization is the strategy the United States is now pursuing.

---

[10] For a discussion of these issues in terms of the U.S. market, see CRS Issue Brief IB10045, *Broadband Internet Access: Background and Issues,* by Lennard G. Kruger and Angele A. Gilroy.

**Table 1. Nippon Telegraph and Telephone's Major Equity
Investments in Overseas Corporations**

Country	Company
U.S.	Verio ($5.5 billion acquisition, pending)
	NTT America (subsidiary)
U.K.	NTT Europe (subsidiary)
Netherlands	Royal KPN NY (15% stake, pending)
France	DoCoMo Europe S.A. (subsidiary)
	NTT France (subsidiary)
Germany	NTT Deutschland (subsidiary)
Australia	Davnet Telecommunications Pty Ltd.
Brazil	NTT do Brasil (subsidiary)
Hong Kong	HIK.Net Co. Ltd. (pending)
	Hutchinson Telecommunications (19% stake, $410 million)
	NTT MSC (subsidiary)
Philippines	Philippine Long Distance Telephone Co.
Sri Lanka	Sri Lanka Telecom Ltd.
Singapore	StarHub Ltd.
	NTT Singapore (subsidiary)
Taiwan	NTT Taiwan (subsidiary)
Thailand	NTT (Thailand) (subsidiary)

Sources: TT on Internet at [http://www.ntt.co.jp]. Guth, Robert A. - NIT Takes on the Big Boys. *Asian Wall Street Journal.* June 5-11. 2000. P. 1. 8.

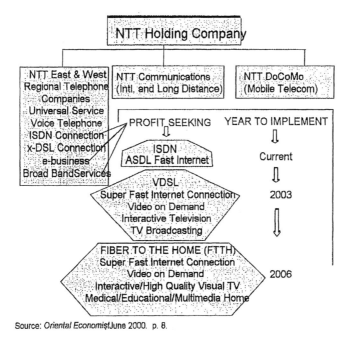

Source: *Oriental Economist* June 2000. p. 8.

Figure 1. NTT's Organization and Plans for Future Internet, TV and Other Services

Chapter 12

REGULATION OF THE TELEMARKETING INDUSTRY: STATE AND NATIONAL DO NOT CALL REGISTRIES

Angie A. Welborn[1]

SUMMARY

Under current federal law, companies that engage in telephone solicitation or telemarketing must maintain a list of consumers who ask not to be called. While there are regulations concerning how such lists are to be maintained and for how long, no federal agency oversees the maintenance of the do not call lists. In addition to the federal regulations, twenty states have laws that create statewide do not call registries that are periodically updated and must be utilized by telemarketers doing business in the state.

In an effort to offer consumers greater protection from intrusive telemarketing calls, legislation has been introduced that directs the Federal Trade Commission to oversee the development of do not call lists in every state, while placing greater restrictions on when telephone solicitation calls can be placed. While this legislation has not yet been enacted, the Federal Trade Commission, acting under the authority of the Telemarketing and Consumer Fraud and Abuse Protection Act, has recently issued a Notice of Proposed Rulemaking that would create a national do not call registry.

This report will discuss current federal regulation of the telephone solicitation industry as well as state laws creating do not call registries. The pending legislation and the FTC's Notice of Proposed Rulemaking will also be discussed. This report will be updated as events warrant.

CURRENT FEDERAL LAW

Both the Federal Communications Commission (FCC) and the Federal Trade Commission (FTC) have promulgated rules that require persons or businesses that engage in

---

[1] Research was provided by LaVonne Mangan, Senio Paralegal Specialist.

telephone solicitations to maintain do not call lists. The current rules do not require the establishment or maintenance of a central nation-wide do not call list.

The Telephone Consumer Protection Act of 1991 directed the Federal Communications Commission to initiate a rulemaking proceeding "concerning the need to protect residential telephone subscribers' privacy rights to avoid receiving telephone solicitations to which they object."[2] The Commission was to develop regulations to implement "the methods and procedures that the Commission determines are most effective and efficient" to accomplish the purposes of the Act. Under the Act, the FCC could have established a "single national database to compile a list of telephone numbers of residential subscribers who object to receiving telephone solicitations, and to make that compiled list and parts thereof available for purchase."[3] However, the FCC chose to require businesses and persons engaged in the telephone solicitation industry to maintain individual do not call lists, rather than establishing a single national list.

Under the FCC's current rules, persons who initiate any telephone solicitation to a residential telephone number must have instituted procedures for "maintaining a list of persons who do not wish to receive telephone solicitations made by or on behalf of that person or entity."[4] The rules also establish minimum standards for maintenance of such lists, including the establishment of a written policy which is to be available on demand, the training of personnel engaged in telephone solicitation, the recording of do not call requests, and disclosure of the identity of the telephone solicitor.[5] Do not call requests must be honored for 10 years from the time the request is made.[6] Calls made to a person with whom the caller has an established business relationship, as well as calls made by or on behalf of a tax-exempt non-profit organization, are exempt from the rules.[7] In addition to enforcement by the FCC, actions for violations of the Act and subsequent rules may be brought by state attorneys general.[8]

The Telemarketing and Consumer Fraud and Abuse Prevention Act directed the Federal Trade Commission to "prescribe rules prohibiting deceptive telemarketing acts or practices and other abusive telemarketing acts or practices."[9] The FTC was instructed to include in the rules "a requirement that telemarketers may not undertake a pattern of unsolicited telephone calls which the reasonable consumer would consider coercive or abusive of such consumer's right to privacy."[10]

In response to this directive, the FTC promulgated the Telemarketing Sales Rule.[11] Under the Telemarketing Sales Rule, it is an abusive telemarketing act or practice for a seller to cause a telemarketer to initiate "an outbound telephone call to a person when that person previously has stated that he or she does not wish to receive an outbound telephone call made by or on behalf of the seller whose goods or services or being offered."[12] The rule includes a safe harbor from liability whereby sellers or telemarketers will not be held liable for

---

[2] 48 U.S.C. 227(c)(1).
[3] 47 U.S.C. 227(c)(3).
[4] 47 CFR 64.1200(e)(2).
[5] *Id.*
[6] *Id.*
[7] 47 CFR 64.1200(f)(3).
[8] 47 U.S.C. 227(f).
[9] 15 U.S.C. 6102(a)(1).
[10] 15 U.S.C. 6102(a)(3)(A).
[11] 16 CFR Part 310.

violations that result from error if they have complied with certain requirements set forth in the rule. They may take advantage of the safe harbor by establishing procedures, training personnel in those procedures, and maintaining a list of persons who have asked not to be called.[13] Certain acts and practices are exempt from the rule, including calls between a telemarketer and any business, unless the calls involve the retail sale of nondurable office or cleaning supplies.[14] The FTC is primarily responsible for enforcing the rule, though actions may also be brought by state attorneys general and private individuals.[15]

STATES LAWS ESTABLISHING DO NOT CALL REGISTRIES

As of January 2002, twenty states have enacted laws to establish statewide do not call registries.[16] The state registries are similar to the lists that telemarketers are required to maintain under current federal laws, but they are generally maintained by a division of the state government, rather than by the telephone solicitation companies themselves. Two states - Maine and Wyoming - do not maintain lists, rather telephone solicitors are required by state law to use the list maintained by the Direct Marketing Association.[17]

Funding for the establishment and maintenance of the lists varies from state to state, with some states requiring consumers to pay a nominal fee to have their telephone number added to the do not call registry. The required fees vary by state. For example, in California the fee cannot exceed one dollar every three years, while in Louisiana the fee is five dollars per year. Most states also require the telemarketers to purchase the do not call list and require payment for periodic updates of the list. Generally, the laws do not allow states to charge more than is required to establish and maintain the list. Fees may be assessed on a sliding scale based upon the size of the telephone solicitation company.

Violations of the do not call laws generally lead to administrative penalties, though in some states consumers may bring private rights of action to recover damages.

---

[12] 16 CFR 310.4(b)(1)(ii).

[13] 16 CFR 310.4(b)(2).

[14] 16 CFR 310.6(g).

[15] 15 U.S.C. 6103 and 6104.

[16] The twenty states are: Alabama, Code of Ala, § 8-19C-2; Alaska, Alaska Stat. §45.50.475; Arkansas, A.C.A. § 4-99-404; California, 2001 Cal. SB 771, to be codified at Cal. Bus. & Prof. Code § 17590; Colorado, 2001 Colo. HB 1405, to be codified at Col. Rev. Stat. § 6-1-901; Connecticut, Conn. Gen. Stat. Ann, § 42-288a; Florida, Fla. Stat. § 501.059; Georgia, O.C.G.A. § 46-5-27; Idaho, Idaho Code § 48-1OO3A; Indiana, 2001 Ind. HEA 1222, to be codified at Ind. Code Ann. § 24.4.7; Kentucky, K.R.S. § 367.46955; Louisiana, 2001 La. FIB 175, to be codified at La. Rev. Stat. 45:844.11; Maine, 32 M.R.S. § 4690-A; Missouri, § 407,1101 R.S.Mo.; New York, NY CLS Gen Bus § 399-z; Oregon, ORS §464.567; Tennessee, Tenn. Code Ann. § 65-4-405; Texas, 2001 Tex. HB 472, to be codified at Tex. Bus. & Coin. Code Ann. § 43.001; Wisconsin, 2001 SB 55, to be codified at Wis. Stat. § 100.52; and Wyoming, Wyo. Stat. § 40-12-302.

[17] The Direct Marketing Association (DMA) is a trade association for telemarketers, telephone solicitation companies, and direct mail companies. The DMA maintains a list of persons who do not wish to receive direct mail advertising or telemarketing calls. Consumers must contact the DMA to be placed on either list. For more information see [http://www.the-dma.org].

PENDING FEDERAL LEGISLATION

On December 20, 2001, Senator Christopher Dodd introduced S. 1881, the Telemarketing Intrusive Practices Act of 2001. The bill requires the Federal Trade Commission (FTC) to "establish and maintain a list for each State, of consumers who request not to receive telephone sales calls; and provide notice to consumers of the establishment of the lists."[18] Under the legislation, the FTC may contract with a state to establish and maintain the lists or may contract with a private vendor to establish and maintain the lists, if such vendor meets certain qualifications.[19] Consumers would notify the Commission of their desire to be included on the list for their state.[20] The Commission would be required to update the lists it maintains "not less than quarterly" and would be required to request state maintained lists annually to "ensure that the lists maintained by the Commission contain the same information contained in the no call lists maintained by individual states."[21] Presumably, this provision would not interfere with state laws providing for do not call registries. The existing registries could simply be used by the Commission for the establishment of the federally required lists so long as the existing lists met the requirements imposed by the federal legislation.

In addition to the lists to be maintained by the FTC, telephone solicitors would be required to maintain a list of consumers who request not to receive calls from that particular solicitor.[22] Telephone solicitors would be required to place requesting consumers on their lists and "provide the consumer with a confirmation number which shall provide confirmation of the request of the consumer to be placed on the no call list of that telephone solicitor."[23] While current regulations require the maintenance of no call lists by telephone solicitors, they do not require the solicitors to provide confirmation of the consumer's request to be placed on the do not call list.

The Federal Trade Commission would be required to make its lists available to telephone solicitors, and the solicitors would be prohibited from making calls to consumers who are on either the Commission's list or on the list maintained by the solicitor.[24] Violations would be pursued by the Federal Trade Commission as an unfair or deceptive trade practice under section 5 of the Federal Trade Commission Act[25] Consumers would also be allowed to bring private rights of action for violations of the do not call provisions and other prohibitions set forth in the legislation, as well as violations of the Federal Trade Commission Act.[26] In a private right of action, a consumer would be able to enjoin the violation; or recover actual monetary loss resulting from the violations, or $500 in damages for each violation, whichever

---

[18] S. 1881, 107th Cong., § 3(a)(2001).
[19] S. 1881, 107th Cong. § 3(b) and (c)(2001).
[20] S. 1881, 107th Cong., § 3(d)(1)(2001).
[21] S. 1881, 107th Cong., § 3(e) (2001).
[22] S. 1881, 107th Cong., § 4(a) (2001).
[23] S. 1881, 107th Cong., § 4 (b) (2001).
[24] S. 1881, 107th Cong., § 5(a)(i) and (2) (2001). The legislation would also prohibit telemarketers from initiating a telephone sales call in the form of an electronically transmitted facsimile or by use of an automated dialing or recorded message service. In addition, telemarketers would be prohibited from making calls between the hours of 9:00 p.m. and 9:00 a.m. and between 5:00 p.m. and 7:00 p.m., local time at the location of the consumer.
[25] S. 1881, 107th Cong., §8(a) (2001). Section 5 of the Federal Trade Commission Act is codified at 15 U.S.C. 45.
[26] S. 1881, 107th Cong., § 7(b)(1)(2001).

is greater.[27] Damages could be tripled where the court finds that the defendant solicitor "willfully or knowingly" violated the provisions of the legislation.[28]

With regard to the pending federal legislation's effect on state law, the bill provides that "nothing in this Act shall be construed to prohibit a State from enacting or enforcing more stringent legislation in the regulation of telephone solicitors."[29] Thus, state laws regulating telemarketers would remain in effect so long as they met or exceeded the federal protections for consumers. As noted above, the FTC would be able to contract with the states to maintain the no call lists, so lists mandated by state law would remain in place.

An earlier bill, H.R. 232, introduced by Representative Peter King, would direct the Federal Trade Commission to promulgate rules requiring telemarketers to "notify consumers who are called that they have the right to be placed on either the Direct Marketing Association's do-not-call list or the appropriate State do-not-call list."[30] Telemarketers would be required to notify either the Direct Marketing Association or the appropriate state of the consumer's request. They would also be required to obtain either the Association's list or the state list on a regular basis.[31] Currently, under federal law, telemarketers are not required to obtain the list maintained by the Direct Marketing Association (DMA), though individual states may require telemarketers to obtain lists maintained by the state or by the DMA.[32]

FTC PROPOSED RULE

On January 22, 2002, the Federal Trade Commission released a Notice of Proposed Rulemaking to amend the Telemarketing Sales Rule to create a national do not call registry.[33] Under the proposed rule, it would be an abusive telemarketing act or practice for a telemarketer to initiate any outbound telephone call to a person who has placed his or her name and/or telephone number on the do not call registry maintained by the Commission.[34] The establishment of a single national registry maintained by the FTC differs from the current regulations which require telemarketing companies to keep their own do not call lists.[35] If the national list is created, telemarketers would be required to purge from their lists the names of persons who are on the FTC's registry.[36]

Telemarketers would be allowed to place calls to persons from whom they have obtained "the express verifiable authorization of such person to place calls to that person."[37] "Express

---

[27] S. 1881, 107th Cong., § 7(b)(1)(B) (2001).

[28] S. 1881, 107th Cong., § 7(b)(2) (2001).

[29] S. 1881, 107th Cong., § 8 (2001).

[30] H.R. 232, 107th Cong., § 2 (2001).

[31] *Id.*

[32] *See supra* note 16.

[33] In addition to creating a national do not call registry, the proposed rule would impose additional disclosure requirements on telemarketers, prohibit the receipt of a consumer's billing information from a third party for use in telemarketing, prohibit the blocking of caller information from caller identification services, prohibit interference with a consumer's right to be placed on a do not call list, and implement provisions of the USA PATRIOT Act that expanded the Telemarketing Sales Rule's coverage to include charitable solicitations. See *[http://www.ftc.gov]* for information on the proposed rule and text of the Notice of Proposed Rulemaking.

[34] Proposed Rule, Sec. 310.4(b)(1)(iii)(B).

[35] The FTC's proposed rule would have no effect on the Federal Communication Commission's regulations regarding telemarketing and company specific do not call lists.

[36] Notice of Proposed Rulemaking, p. 66.

[37] Proposed Rule, Sec. 310.4(b)(1)(iii)(B)(1) and (2).

verifiable authorization" could be obtained in one of two ways: (1) written authorization including the consumer's signature; or (2) oral authorization that is recorded and authenticated by the telemarketer as being made from the telephone number to which the consumer is authorizing access.[38] According to the FTC, this would "provide consumers with a wider range of choices" than the current regulations provide by allowing the consumer to decide if he or she wants to eliminate *all* telemarketing calls covered by the rule, eliminate calls only from specific telemarketers through inclusion on lists maintained by individual companies, or expressly authorize calls from specific organizations or sellers.[39]

The Notice of Proposed Rulemaking does not address jurisdictional issues regarding the enforcement of the proposed national do not call registry in states that have enacted laws establishing their own do not call registries. The Commission has requested comments on "the interplay between the national registry and State 'do-not-call' schemes."[40]

The Federal Trade Commission will be accepting comments on the proposed rule until March 29, 2002.

---

[38] Proposed rule, Sec. 310.4(b)(1)(iii)(B)(1) and (2).
[39] Notice of Proposed Rulemaking, p. 67.
[40] *Id.* At 72.

Chapter 13

ENCRYPTION TECHNOLOGY: CONGRESSIONAL ISSUES

Richard M. Nunno

SUMMARY

The controversy over encryption concerns what access the government should have to encrypted stored computer data or electronic communications (voice and data, wired and wireless) for law enforcement and national security purposes.

Encryption and decryption are methods of using cryptography to protect the confidentiality of data and communications. When encrypted, a message only can be understood by someone with the key to decrypt it. Businesses and consumers want strong encryption products to protect their information, while the Clinton Administration wants to ensure the law enforcement community's ability to monitor undesirable activity in the digital age.

Until recently, the Administration promoted the use of strong encryption (greater than 56 bits) here and abroad, only if it had "key recovery" features where a "key recovery agent" holds a "spare key" to decrypt the information. The Administration wanted key recovery agents to make the decryption key available to authorized federal and state government entities. Privacy advocates argued that law enforcement entities would have too much access to private information.

Under this policy, the Administration was using the export control process to influence whether companies develop key recovery encryption products by making it easy to export products with key recovery, and difficult for those products without. There were no limits on domestic use or import of any type of encryption, so the Administration tried to influence what is available for domestic use through export controls since most companies do not want to create two products—one for U.S. use and another for export. In September 1997, however, FBI Director Louis Freeh raised the possibility of requiring encryption products manufactured in or imported into the United States to have key recovery features and opened the possibility that key recovery could be enabled by the manufacturer, not only the user. U.S. companies argued that U.S. export policies hurt their market share while helping foreign companies that are not subject to export restrictions. Many businesses and consumer groups agree that key recovery is desirable when keys are lost, stolen, or corrupted, but want market forces to drive the development of key recovery encryption products. They also object to government having any role in determining who can hold the keys.

All parties agreed that encryption is essential to the growth of electronic commerce and use of the Internet, but until September 1999, industry and privacy rights groups opposed the Administration's policy. In the 106th Congress, legislation was introduced intended to foster widespread use of the strongest encryption (H.R. 850, S. 798). While the Administration continued to oppose that legislation, H.R. 850 was marked up by five Committees, resulting in widely varying and, in places, opposing legislation. S. 798 passed the Senate Commerce Committee and awaits further action.

In September 1999, the Administration announced plans to further relax its encryption export policy, and the rules for implementing that policy were issued by the Department of Commerce on January 14, 2000, after which, pressure from the industry (but not from privacy rights groups) to enact encryption legislation subsided.

MOST RECENT DEVELOPMENTS

On October 2, the National Institute for Standards and Technology announced its selection of a new Advanced Encryption Standard for federal agency use to protect unclassified information. The algorithm for the standard, known as Rijndael, was developed by Belgian scientists and will be made freely available for use by the private sector.

On July 17, the Administration announced further updates to its encryption policy to allow exports of any strength encryption products to the governments of European Union and eight additional countries. It also allows exports, without a technical review or reporting requirements, of encryption embedded in short-range wireless products (e.g., mobile phones, audio devices, cameras and videos).

BACKGROUND AND ANALYSIS

Encryption, Computers, and Electronic Communications

Encryption and decryption are procedures for applying the science of cryptography to ensure the confidentiality of messages. Technically, the issue discussed here is cryptography policy, but since encryption is the most controversial application of cryptography, it is the term used popularly and herein. (There are other methods of using cryptography to protect confidentiality — steganography and "chaffing and winnowing" — but constraints on the length of this issue brief do not permit discussion of them.)

Encrypting messages so they can be understood only by the intended recipient historically was the province of those protecting military secrets. The burgeoning use of computers and computer networks, including the Internet, now has focused attention on its value to a much broader segment of society. Government agencies seeking to protect data stored in their databases, businesses wanting to guard proprietary data, and consumers expecting electronic mail to be as private as first class mail, all want access to strong encryption products. Other users of electronic communications, for example cellular (wireless) phone users who expect calls to be as private as wireline calls, also are showing increased interest in encryption. While encryption is uncommon for telephone users today, the advent of digital telephone services (particularly Personal Communication Services, PCS, a digital form of cellular telephony) is expected to make encrypted voice and data communications over telephones more common.

Whether hardware- or software-based, an encryption product scrambles a message using mathematical algorithms. A corresponding key is needed to decrypt (unscramble) the message, and the key itself also may be encrypted. The algorithm is a series of digital numbers (bits), and the level of difficulty of breaking the code (its "strength") is usually represented by the number of bits in the key. (There are other factors that affect a key's strength, but in this debate, bit length is used as a benchmark.) Unencrypted data are referred to as "plaintext." Encrypted data are "ciphertext."

The National Institute of Standards and Technology (NIST), in conjunction with industry, developed an encryption standard using a 56-bit key in 1977. Called the Data Encryption Standard (DES), it is widely used today in the United States and abroad, often in an enhanced mode called "3-key triple DES" providing the equivalent of a 112- bit key. NIST is currently working to establish a new, stronger standard than DES referred to as the Advanced Encryption Standard (AES). The need for a stronger standard was underscored in 1997 when DES was broken (see below).

Encryption products are widely available today, including some that use 128-bit keys or more. Some 128-bit encryption software can be downloaded from the Internet. There are no limits on the strength of encryption products used in the United States, whether acquired here or imported. The only limits are on exports. This indirectly influences what is available domestically, however, since most U.S. companies are reluctant to develop two products, one for the U.S. market and another for export. For many years, reflecting the policies of the past three Administrations, the State Department did not allow general exports of encryption with more than 40-bit keys, except for banking and U.S. -owned subsidiaries.

In December 1996, the Clinton Administration raised the limit to 56 bits for easily exportable encryption products that do not have key recovery features (see box), and removed bit length limits entirely for products with key recovery. Breaking a message encrypted with a 40-bit key by "brute-force" (trying every possible combination of bits until the correct one is found) is not considered difficult. In January 1997, a 40-bit key was broken in 3.5 hours. In general, for each bit added to the encryption key, the time required to break the key doubles. Thus it takes 216 (65,536) times longer to break a 56-bit key than a 40-bit key. In July 1998, a group from the Electronic Freedom Foundation (an electronic privacy rights group) demonstrated (for less than $250,000) the vulnerability of 40- and 56-bit keys by using a network of nearly 100,000 PCs on the Internet that broke a 56-bit key in 22 hours, 15 minutes. The ability to break 128-bit encryption (considered strong encryption) has not yet been demonstrated in the commercial sector.

KEY RECOVERY AND KEY RECOVERY AGENTS

Once called "key escrow," key recovery means that when stored data or electronic communications are encrypted, a third party has a copy of the key needed to decrypt the information. The third party is called a key recovery agent (formerly a key escrow agent). Key recovery is widely regarded as useful in cases where a key is lost, stolen or corrupted. Most parties to the encryption debate agree that market forces will drive the development of key recovery-based encryption products for stored computer data because businesses and individuals will want to be sure they can get copies of keys in an emergency. It is less clear if market demands will drive key recovery systems for electronic communications.

The controversy is over governments attempt to "encourage" the development of key recovery-based products through the export control process, the governments role in determining who can serve as key recovery agents. and the extent to which law enforcement agencies could obtain the key if they suspect undesirable activity (terrorism, child pornography, and drug cartels are often cited as examples).

In May 1996, the National Research Council (NRC) released a comprehensive report entitled *Cryptography's Role in Securing the Information Society* (the "CRISIS" report). It stressed that national policy should make cryptography broadly available to all legitimate elements of society, promote continued economic growth and leadership of key U.S. industries, and ensure public safety and protection against foreign and domestic threats. Among the recommendations: key escrow is an unproven technology and the government should experiment with it and work with other nations, but not aggressively promote it now; export controls should be relaxed progressively, but not eliminated; and encryption policy issues can be debated adequately in public without relying upon classified information. The report also recommended that no law should bar the manufacture, sale or use of any form of encryption within the United States; and government should promote information security in the private sector. The report underscored that utilization of strong encryption and law enforcement objectives can be mutually compatible.

Business and consumer groups consider 56-bit keys inadequate to ensure privacy and security. They oppose export encryption controls and requirements for key recovery features. They object to the government using the export process to force the development of key recovery encryption products, rather than allowing market forces to prevail, and to the government's role in determining who can serve as key recovery agents. These groups argue that strong encryption is needed, for example, to enhance the prospects for electronic commerce and other uses of computer networks. The willingness of consumers to buy goods via the Internet could be markedly affected by their beliefs as to whether credit card numbers will be secure. Businesses using computers for either internal or external communications need to ensure that competitors or other unauthorized parties cannot gain access to proprietary information. Privacy advocates argue that consumers should be assured that personal, medical and financial information transmitted by or stored in computers will be protected. They note that since 128-bit non-key recovery encryption is available worldwide either by downloading it from the Internet or buying it from foreign firms, the U.S. government already has lost control of influencing its availability. A 1997 survey conducted by Trusted Information Systems found 656 foreign encryption products available from 29 countries (in addition to 963 U.S. products). A 1998 report by the Economic Strategy Institute, *Finding the Key,* concluded that if the Administration's current policies remained in effect, the U.S. economy would lose $35-96 billion by 2002 in lost encryption product sales; slower growth in encryption-dependent industries; foregone cost savings and efficiency gains from the Internet, intranets, and extranets; and indirect effects throughout the economy.

In June 1999, a study conducted by George Washington University, titled the "Growing Development of Foreign Encryption Products in the Face of U.S. Export Regulations," found over 800 encryption products available from 35 foreign countries [http://www.seas.gwu.edu/seas/institutes/cpi]. At least 167 of those products were found to use strong encryption. Over 512 companies either manufacture or distribute cryptographic products (of quality comparable to U.S. products) in over 70 countries outside the United States. In June 1999, the Electronic Privacy Information Center (EPIC) produced its second annual report on the encryption policies of foreign nations and international organizations titled "Cryptography and Liberty," concluding that "in the vast majority of countries [both developed and developing] cryptography may be used, manufactured, and sold without restriction." On April 3, 2000, EPJC released its third annual international review of encryption policies, covering 135 countries. The report found the relaxation of export controls was continuing, but that law enforcement agencies were seeking new authority and new funding to gain access to private keys and personal communications.

Supporters of encryption export controls agree that strong encryption is needed but insist that law enforcement and national security concerns demand that, when authorized, the government be able to intercept and decrypt electronic communications or decrypt stored computer data where undesirable activity is suspected. Law enforcement and national security officials want to ensure their ability to access the plaintext of the information. The method most often discussed is to obtain the key needed to unscramble encrypted information from key recovery agents. Hence, they support the use of strong encryption products as long as they include key recovery features, and want to limit the development of strong non-key recovery products. While conceding some strong non-key recovery encryption products already are available, they claim use of these products is not widespread. They argue that while the U.S. government cannot prevent the availability of strong non-key recovery encryption, at least it can be restrained, and future generations of encryption products (with key recovery) will displace those now in use.

Although the publicity surrounding the encryption debate so far has centered on access to stored computer data, electronic communications are equally important to the law enforcement and national security communities. An Internet message, for example, is stored data when it resides on a server or an individual's computer, but it is an electronic communication while it is being transmitted between computers. The encryption export regulations apply to products for encrypting other electronic communications, not just those between computers. Telephones, whether wired or wireless (such as cellular phones), are also covered, for example. The 1994 Communications Assistance to Law Enforcement Act (CALEA, often called the "Digital Wiretap" Act, P.L. 103-414) requires telecommunications carriers to ensure their equipment permits the interception of any electronic communication by law enforcement officials. If the communication is encrypted, law enforcement agencies want to ensure they can decrypt it, too. (CALEA requires the telecommunications carrier to provide decrypted information if the carrier itself is responsible for the encryption, hut not if the customer has encrypted it.)

PROPONENTS AND OPPONENTS OF ENCRYPTION EXPORT CONTROLS

Proponents include the Clinton Administration (notably the Department of Justice and the National Security Agency) and others who are concerned about the ability of terrorists and other criminal groups to conduct activities unmonitored if strong non-key recovery encryption is widely available.
Opponents include:

• computer hardware and software manufacturers who do not want to develop separate products for domestic and foreign markets and worry they will lose market share to foreign competitors who do not have to abide by such limits. They also are concerned that no one may buy encryption products for which the U.S. government can obtain the key.

• U.S. businesses that want to use the same computer systems they have in their home offices with their foreign clients; and

• privacy and consumer groups who want individuals to have access to the best encryption possible without regard to key recovery features.

ADMINISTRATION POLICY: GRADUAL RELAXATION OF RULES

The Clinton Administration has always strongly supported arguments by law enforcement and national security agencies that the government must be able to gain access to the plaintext of encrypted electronic data and messages when undesirable activity is suspected. In addition to international criminal activity, the Administration (notably the FBI) wants to be able to monitor domestic criminal activity. The Administration has always permitted use of any strength encryption, without a key recovery requirement, in the United States. Rather than attempting to change that policy directly, the Administration used the indirect route of export controls to influence what types of encryption products were available, both here and abroad.

Initially, the Clinton Administration sought to restrain the development of strong encryption products by prohibiting export of greater than 40-bit encryption (with a few exceptions). The Administration also tried several approaches to promote "voluntary" use of key recovery agents. In April 1993, the Administration released its "Clipper chip" policy-requiring emplacement of special encryption computer chips (called Clipper, an encryption device used for unclassified but sensitive government communications) into new government equipment for voice communications, with two government agencies, NIST and the Department of Treasury, jointly serving as key escrow agents (each holds part of the key). The Administration implemented this policy in 1994 for sensitive but unclassified voice communications in the federal government through a Federal Information Processing Standard (FIPS) called the Escrowed Encryption Standard (EES, or FIPS-185). The Administration hoped that industry would accept the Clipper chip for its own use, but industry strongly objected to the key escrow provisions, particularly the fact that government agencies would hold the keys. In July 1994, the Administration agreed to work with the private sector to develop a "voluntary" key escrow system for data using "trusted third parties" as escrow agents instead of government agencies. This proposal was referred to by its detractors as "Clipper II."

Industry continued to object to the key escrow concept as well as the export controls, leading to the legislation discussed below. In May 1996, the Administration released a draft paper on encryption policy, followed by a July statement by Vice President Gore. Called "Clipper III" by its opponents, these documents outlined policy changes the Administration was considering. Among other things, the term "key recovery" replaced "key escrow" to emphasize the positive attributes of key escrow in providing a means to recover a key that is lost, stolen, or corrupted. Furthermore, "key escrow" had come to be identified with the concept of the government holding the key. Under the key recovery policy, a trusted third party or an organization itself can serve that function ("self escrow") with some restrictions.

On October 1, 1996, Vice President Gore announced the changes to the Administration's policy, to focus on the need for strong encryption, as long as it included key recovery features. The key recovery agent would be required to give the key to duly authorized law enforcement officials if undesirable activity is suspected (the three types most often cited are drug cartels, child pornographers, and terrorists). The associated executive order was signed November 15, and the Administration published an "interim final" regulation on December 30, to last two years, with the following details: continuation of no restrictions on domestic use or import of any encryption; no key length restrictions on export of encryption products if a key recovery system was used for that product; for 56-bit encryption products without key recovery systems, a one-time review was required before exporting the product, and within two years the exporter must have developed a key recovery system; export licenses were granted in 6-month increments to hold exporters to a timetable to ensure the key recovery systems were being developed (if not, the export license was not renewed); trusted parties served as key recovery agents, and in some cases, organizations were allowed to escrow keys themselves (self-escrow) if they met certain requirements; and commercial encryption was removed from the Munitions List and responsibility for commercial encryption export licensing was transferred from the State Department to the Commerce Department, with the Department of Justice serving an advisory role in commercial encryption export decisions. Foreign governments would apply to U.S. courts to gain access to keys, as they do when seeking other types of evidence. During the next two years, the Commerce Department granted some waivers for banking and financial services.

On September 3, 1997, FBI Director Louis Freeh testified to the Senate Judiciary Committee that there was a need for domestic use restrictions on encryption products. Following that testimony, the existence of legislation proposed by the Administration to impose domestic use restrictions became widely known among congressional and industry groups. Vice President Gore later stated that Freeh's comments reflected the FBI's view, but did not indicate a change in Administration policy. The House Intelligence Committee, however, later approved an amendment to legislation similar to Freeh's position about the need for key recovery to be built into encryption products.

In March 1998, Vice President Gore wrote to Senator Daschle restating the Administration's desire for a "balanced approach" to encryption policy and seeking to "produce cooperative solutions, rather than seeking to legislate domestic controls." The discussions would also enable additional steps to relax export controls on encryption products. On April 15, Secretary of Commerce Daley announced the release of a new report on electronic commerce wherein lie said that although the Administration's policy was the right one, its implementation was a failure. He urged both industry and government to strive harder to reach consensus on the issue. In April 1998, Undersecretary of Commerce for the Bureau of Export Administration (BXA) Reinsch commented at a Congressional Internet Caucus meeting that the Administration was no longer looking for a legislative solution to the encryption issue.

In July 1998, the Administration declassified two security algorithms used in the Clipper chip. In September 1998 it announced plans to allow the export of 56-bit encryption products without requiring provisions for key recovery, after a one-time review, to all users outside seven "terrorist countries" (Cuba, Iran, Iraq, Libya, North Korea, Syria, and Sudan). Export of encryption products of any strength was permitted to 46 designated countries if key recovery or access to plaintext is provided to a third party. The Administration also supported an FBI proposal to establish a technical support center to help law enforcement keep abreast of encryption technologies. In December 1998, the BXA released interim rules to implement the Administration's export control policy initiative. The rules allowed for the export of encryption commodities and software to U.S. companies or subsidiaries in the finance, insurance, health-care and medical end-users, and electronic commerce industries.

In September 1999, the Administration announced its new encryption export policy, to make encryption products of any key length, after a technical review, exportable without a license to users in any country except seven "terrorist countries". Regulations implementing the Administration's new encryption export policy were issued by the BXA on January 14, 2000. According to the new rules, retail encryption commodities and software of any key length can be exported without a license to any non-government end user in any country except the seven state supporters of terrorism, and can be re-exported to anyone (including Internet and telecommunications service providers). Exports previously allowed only for a company's internal use can now be used for communication with other firms, supply chains, and customers. Exports to government end-users still require a license. Exporters must report to BXA on where the encryption product is exported, and BXAw1li determine whether products qualify as retail by reviewing their functionality, sales volume, and distribution methods. In addition, if source code (computer language instructions written by programmers) is made publicly available (under "open source" policies) and no royalty is charged for its use, the code is not subject to export restrictions, a provision that was not included in an earlier draft and was the source of criticism by industry and privacy rights groups.

Industry Reactions

Most participants in the debate agreed that market forces will lead to the development of key recovery-based encryption products for stored data because companies and individuals will want to be able to replace lost, stolen or corrupted keys. The debate was over the government's role in "encouraging" the development of key recovery products through export regulations and the access government has to the keys. Also of concern was the government's inclusion of other electronic communications.

While computer companies continued to argue against the Administration's policy, some also developed key recovery products to satisfy the policy. A group of companies formed the

Key Recovery Alliance [http://www.kra.org] in 1996 to identify barriers to the development of marketable key recovery products. On March 4, 1998, a group of more than 100 companies and organizations (some also members of the Key Recovery Alliance) formed the Americans for Computer Privacy (ACP) coalition [http://xvww.computerprivacy.org] to lobby for enabling the use of strong encryption, against export controls on strong encryption, and against mandatory key recovery. Among the members are America On Line, Microsoft, Sybase, the National Rifle Association, the Law Enforcement Alliance of America. and the Business Software Alliance.

In March 1998, Network Associates announced that it had arranged for a Swiss company to develop its own software product using specifications in a book by Philip Zimmerman, the creator of Pretty Good Privacy (PGP). Network Associates bought Mr. Zimmerman's company in 1997. PGP is a 128-bit encryption product that does not have key recovery and hence could not be exported under the current regulations. (An older version is available via the Internet, however, which sparked a multi-year Justice Department investigation of Mr. Zimmerman that resulted in no action against him). Since the book may be exported, and the Swiss company received no technical assistance from Network Associates, the company believes no laws were broken. The Swiss product was sold by a Dutch firm under the PGP name. Opponents of encryption controls pointed to this as evidence that the U.S. government cannot control the spread of non-key recovery encryption.

In July 1998, as a possible compromise to the Administration, a group of software companies announced plans to develop a product, "private doorbell", to capture data that could be given to law enforcement before it is encrypted and sent over the Internet. While industry groups approved of the Administration's encryption export policy, they argued that 56-bit encryption had been broken and that stronger encryption was necessary. Furthermore, the Department of Commerce's rules could render the policy ineffective in increasing their ability to export encryption products. Privacy rights groups argued that while the new policy may help big companies, it would not increase the availability or use of 56-bit or stronger encryption by individual users of Internet communications.

While the computer industry is satisfied with the latest rules, some privacy rights groups (including the American Civil Liberties Union, the Electronic Frontier Foundation, and the Electronic Privacy Information Center) argue that the remaining ambiguities in the rules make encryption technology overly cumbersome for individuals to use. Because the regulations could be reversed by a future Administration, these groups still advocate the passage of legislation to codify the changes in U.S. encryption policy.

ACTION IN THE 105TH CONGRESS

Three bills in the House and four in the Senate concerning encryption were introduced in the 105th Congress; none were enacted, although one (H.R. 1903) passed the House. Three (H.R. 695, S. 376, and S. 377) were versions of bills considered in the 104th Congress, generally favoring relaxed encryption export controls. Four new bills were also introduced. S. 909 reflected a philosophy closer to that of the Clinton Administration than the three previous bills. H.R. 1903 focused broadly on computer security issues and the role of NIST H.R. 1964 focused broadly on computer privacy and security issues. S. 2067 was generally viewed as pro-industry and pro-privacy. Hearings were held by six House committees (Commerce, International Relations, Intelligence, Judiciary, National Security, and Science) and two Senate committees (Judiciary; and Commerce, Science, and Transportation).

The Security and Freedom Through Encryption (SAFE) Act (H.R. 695, Goodlatte), as introduced, sought to relax export controls on encryption, although versions of H.R. 695 as reported from various committees had substantially different provisions. The bill was eventually considered by the Committees on Judiciary, International Relations, Commerce, National Security, and Intelligence. Amendments adopted by the latter two committees reversed some of the provisions of the original bill by maintaining or increasing restrictions on export controls of encryption. The Rules Committee did not take action on the bills.

The Encrypted Communication Privacy Act of 1997 (S. 376, Leahy) prohibited mandatory use of key recovery but allowed law enforcement to access the key under court order if key recovery is used; codified existing domestic use policy; gave the Secretary of Commerce exclusive jurisdiction over commercial encryption exports; liberalized export controls; made it a crime to use encryption to obstruct justice; and established liability protection and penalties for "key holders." The bill also established procedures for foreign governments to access keys or decryption assistance.

The Promotion of Commerce On-Line in the Digital Era (PRO-CODE) Act of 1997 (S. 377, Bums) prohibited mandatory key recovery and established an Information Security Board as a forum to foster communication and coordination between industry and government. The bill also codified existing domestic use policy and gives the Secretary of Commerce exclusive jurisdiction over commercial encryption exports. It liberalized export controls but required the publisher or manufacturer of encryption software or hardware to report to the Secretary of Commerce within 30 days after exporting a product on the product's encryption capabilities. The report would have included the same information required under the December 30, 1996 regulations, but would be provided after export instead of as a condition of obtaining a license.

The Secure Public Networks Act (S. 909, McCain et. al.) codified existing domestic use policy; established penalties for use of encryption in commission of a crime; encouraged but did not require use of key recovery; established procedures for government approval of key recovery agents and certificate authorities; required key recovery agents, whether or not registered by the government, to disclose recovery information to lawfully authorized federal or state government entities; provided liability protection for key recovery agents acting pursuant to the Act; permitted export of 56-bit encryption products without key recovery if they meet certain conditions; permitted export of any strength encryption product if it is based on a qualified key recovery system and meets certain other conditions; and allowed the President to waive provisions of the bill for national security or domestic safety and security reasons. During and after markup, the committee adopted amendments establishing an Encryption Export Advisory Board (EEAB) with four government (CIA, FBI, NSA and the White House) and eight industry representatives to make recommendations on whether export exemptions should be granted for non-key recovery products stronger than 56-bits. The committee ordered the bill reported, but the report was never filed.

The Senate Judiciary Committee requested sequential referral of S. 909, and held a hearing on July 9, 1997 where FBI Director Louis Freeh expressed reservations about the bill because it allowed widespread use of strong encryption within the United States regardless of whether it has key recovery. Freeh amplified his concerns at a September 3 hearing before the

Subcommittee on Technology, Terrorism, and Government Information. He stated that he wanted U.S. manufacturers to be required to build key recovery into encryption products, and that imported encryption products also be required to include key recovery. He further stated that achieving the goal of immediate lawful decryption "could be done in a mandatory manner. It could be done in an involuntary manner. ..." He later added that he thought legislation should first include the requirement that key recovery be built into encryption products and "then take up the more complex discussion about how that's enabled...." He also stated that Internet service providers should be required to be able to decrypt communications immediately.

The Encryption Protects the Rights of Individuals from Violation and Abuse in Cyberspace (E-PRIVACY) Act (S. 2067, Ashcroft, et. al.) prohibited federal or state agencies from linking the use of encryption for authentication or identification to the use of encryption for confidentiality purposes; required that the use of encryption products be voluntary and market-driven; outlined procedures for U.S. and foreign law enforcement agencies to access decryption keys or assistance in decrypting electronic communications or stored data; established a National Electronic Technologies (NET) Center in the Department of Justice to help law enforcement keep pace with encryption technology; made the use of encryption to obstruct justice a crime; established an Encryption Export Advisory Board to determine whether comparable foreign products are commercially available; maintained the President's authority to prohibit export of encryption products to countries that support international terrorism or to impose embargoes on exports to or imports from a specific country; made electronic records in networked storage be treated in law as though the record had remained in the possession of the person who created the record; and set the circumstances under which the government may require a mobile electronic communication service to reveal the real-time physical location of a subscriber, and may obtain information from pen register and trap and trace devices.

The Computer Security Enhancement Act (H.R. 1903, Sensenbrenner) amended and updated the Computer Security Act of 1987, enhancing the role of the National Institute of Standards and Technology (NLST). As passed by the House (H.Rept. 105-243), the bill required NIST to promote the use of commercial-off-the-shelf encryption products by civilian government agencies; clarified that NIST standards and guidelines are not intended as restrictions on the production or use of encryption by the private sector; provided funding for computer security fellowships at NIST; and required the National Research Council to conduct a study of public key infrastructures. A section requiring NIST to develop standardized tests and procedures to evaluate the strength of foreign encryption products was removed before passage. The bill passed the House and was reported without amendment by the Senate Commerce Committee October 13, 1998 (S.Rept. 105-412).

The Communications Privacy and Consumer Empowerment Act (H.R. 1964, Markey) covered a range of computer privacy and security issues. With regard to encryption, the bill codified existing domestic use policy, prohibited the government (federal or state) from conditioning the issuance of certificates of authentication or certificates of authority upon use of key recovery systems, and prohibited the government (federal or state) from establishing a licensing, labeling or other regulatory scheme that requires key escrow as a condition of licensing or regulatory approval. The bill also required the National Telecommunications and Information Administration (NTIA) to conduct a study on, inter alia, how data security issues affect electronic commerce, including identification of generally available technologies (including encryption) for improving data security.

ACTION IN THE 106TH CONGRESS

On February 25, 1999, Representative Goodlatte introduced a new version of SAFE (H.R. 850), similar to the bill introduced in the 105th Congress, with a new provision directing the Attorney General to compile examples in which encryption has interfered with law enforcement. The bill was reported by the Judiciary Committee on April 27, and was referred jointly and sequentially to Committees on International Relations, Commerce, Armed Services, and Intelligence. Hearings were held by the Committees on Commerce (May 5 and 25), International Relations (May 18) and Intelligence (June 9 and July 14), Armed Services (July 1 and 13). The bill has gained 257 co-sponsors, with a majority of both Republican and Democratic leadership. The bill was reported (amended) by the Committee on Commerce, (H.Rept. 106-117, Part II) on July 2, and by the remaining three committees (parts III, IV, and V) on July 23.

The five versions of H.R. 850 differ significantly. The versions passed by the Committees on the Judiciary, Commerce, and International Relations codify the regulations for unrestricted domestic use and sale of encryption, prohibit the government from mandating key escrow practices for the public, and liberalize the controls governing the export of strong encryption. One of the amendments passed by the Commerce Committee made it a crime to fail to decrypt information upon court orders, raising opposition from privacy rights advocates. The Armed Services and Intelligence Committee versions, in contrast, have minimal or no mention of domestic use of encryption, and increase the authority of the President in restricting the controls governing the export of strong encryption. All of the bills, except for the version by the Armed Services Committee, establish criminal penalties for the use of encryption in the furtherance of a criminal act. The Intelligence Committee bill, however, provides greater details than the others for criminalizing the use of encryption in a criminal act. In addition, each Committee added provisions for specific agencies and circumstances. For example, the Commerce Committee established a National Electronic Technologies (NET) Center in the Department of Commerce to promote the exchange of information regarding data security techniques and technologies, and the International Relations Committee directed the Secretary of Commerce to consult with the Attorney General, the Federal Bureau of Investigation, and the Drug Enforcement Administration before approving any license to export encryption products to any country identified as being a major drug producer. The Intelligence Committee bill authorizes appropriations for the Technical Support Center, at the FBI. The bill was sent to the House Rules Committee for further action on July 23, 1999.

In the Senate, On April 14, 1999, Senators McCain, Burns, Wyden, and Leahy introduced S. 798, the Promote Reliable Transactions to Encourage Commerce and Trade (PROTECT) Act. The bill would immediately raise the maximum exportable key length of encryption products to 64 bits; S. 798 also sets a deadline of January 1, 2002 for the federal adoption of the Advanced Encryption Standard (which uses a 128 bit key length) and allows the export of products employing AES at that date. S.798 allows the export of strong (greater than 64 bit) encryption products with key recovery features, as well as the export of strong encryption products to "legitimate and responsible entities," including publicly traded firms, U.S. corporate subsidiaries or affiliates, firms required by law to maintain plaintext records, and others. 8.798 does not contain criminal provisions for the use of encryption in the furtherance of a crime (unlike H.R.850), and prohibits domestic controls and mandatory plaintext access. The Senate Commerce Committee held a hearing on the bill on June 10. The Senate Commerce Committee approved S. 798 on June 23, 1999.

Deep divisions remain between those who oppose liberalizing encryption policy and those who advocate it. While the computer industry, and privacy and consumer advocacy groups generally favor both bills, H.R. 850 is considered more pro-industry because of its greater liberalization of encryption export controls. The Administration remains opposed to both bills, arguing that the current export restrictions are necessary to prevent the use of encryption by undesirable groups, and has increased its efforts to maintain the status quo on encryption policy. Reaching a compromise on some of the differences (such as key escrow and export policies) may be a difficult task. In addition, on July 27, 1999, Representative Goss introduced two more bills on encryption policy: H.R. 2616 (which reflects House Intelligence Committee mark-up of H.R. 850), and H.R. 2617 (which proposes tax incentives for the nations encryption software manufacturers to develop products with key recovery). The prospects for enacting legislation are further complicated by a possible veto by President Clinton if the final bill passed by Congress is not supported by officials in the Defense and Justice Departments.

Three other bills have been introduced containing provisions regarding encryption. On April 21, 1999, Senator Leahy introduced S. 854, which includes provisions that promote the use of encryption by (1) prohibiting government requirements for non-federal use of key recovery or plaintext access practices, (2) prohibiting federal agencies from requiring non-federal entities to use specific encryption products to receive services, (3) prohibiting federal encryption products that interact with commercial systems from interfering with the encryption capabilities of the commercial products, and (4) prohibiting the disclosure of decryption assistance to foreign governments without a Court order. The bill was referred to the Committee on the Judiciary. On June 9, 1999, Representative Sensenbrenner introduced H.R. 2086, which includes a provision directing the National Science Foundation to undertake a study comparing the availability of encryption technologies in foreign countries to those subject to U.S. export restrictions. On July 1, 1999, Representative Sensenbrenner introduced H.R. 2413, which contains provisions for establishment of a public key management infrastructure and encryption standards for federal computers.

On September 21, 1999, the President sent to Congress proposed legislation that would ensure that law enforcement maintains its ability to access decryption information stored with third parties, and is allowed to withhold information in courtrooms on its techniques used in decryption. The bill would also authorize $80 million over four years for the FBI Technical Support Center, which will serve as a technical resource in responding to the use of encryption by criminals. To date, no Member has introduced that legislation.

ISSUES

Key Recovery

Key recovery is the fundamental tenet of the Clinton Administration policy. The Administration wants law enforcement access to keys for encrypted data stored by computers, transmitted between computers, or other types of electronic communications. Not only does the Administration view this as critical for U.S. users, but it seeks creation of a global key management infrastructure (KMI, now referred to as public key infrastructure, or PKI) to ensure confidentiality for the growth of global electronic commerce, and monitoring undesirable activity (by terrorists, drug cartels, or child pornographers, for example).

Many opponents of encryption controls agree that key recovery has advantages for recovering a lost, stolen, or corrupted key, but believe market forces will drive the development of a KMI for stored computer data without government involvement. Less likely is a market-driven demand for key recovery products for electronic communications. In any case, opponents of controls insist that the government should have no role in choosing who holds the keys. They fear the government will have unfettered access to private files and communications, though the Clinton Administration stresses that proper legal authorization will be required. Liability protection for proper release of keys, and penalties for improper use or release of keys, is an important aspect of Administration policy.

Questions about technical vulnerabilities that could be introduced if key recovery is incorporated into computer systems were raised in a report (updated June 1998), *The Risks of Key Recovery, Key Escrow, and Trusted Third Party Encryption,* by an ad hoc group of cryptographers and computer scientists. They concluded that key recovery "introduces a new and vulnerable path to the unauthorized recovery of data" and the "massive deployment of key-recovery-based infrastructures to meet law enforcement's specifications will require significant sacrifices in security and convenience and substantially increased costs...."

The Administration acknowledges that global agreement on key recovery and KMI policy is essential to its policy and has been working with the Organization for Economic Cooperation and Development (OECD) to develop guidelines for a global KML In 1997, the OECD released those guidelines, which state that "national cryptography policies *may* allow lawful access to plaintext, or cryptographic keys, or encrypted data" (emphasis added). Hence, OECD neither endorsed nor rejected the concept of law enforcement access to decryption keys. The European Commission published a communication in October 1997 that noted the need for strong encryption to advance electronic commerce and expressed strong reservations about regulating encryption (by requiring key recovery, for example). Since then, Canada, Finland, Germany, France, and Taiwan have announced a relaxation or elimination of their key recovery laws. Lacking international consensus, many believe it is unlikely that mandated key recovery will survive.

Export Restrictions

Using the export process to influence the type of encryption products that are available in the United States and abroad is one strategy of the Administration's policy. The

Administration points to threats to national security and public safety that would arise if criminals and terrorists used encryption that the U.S. government could not decrypt. Administration representatives argue that NSA, for example, has been able to thwart criminals and terrorists because NSA intercepted communications in time; if those communications had been encrypted with strong encryption, their task would have been much more difficult. NSA opposes passing a law that does not require companies to notify the government of what encryption products are being exported and to whom. Others point to difficulties in stopping future attacks in an era when terrorists could use strong encryption.

Opponents of the Administration's policy counter that the United States, through export controls, cannot prevent access to strong non-key recovery encryption by criminals and terrorists because it is already available elsewhere in the world. They further point out that the current policy of no restrictions on domestic use or import of encryption means that domestic threats would not be affected.

Until its September 1999 announcement, the Administration was using the export process to encourage companies to develop products with key recovery features. There were no limits on the strength of encryption products that can be exported if they include key recovery. Opponents of export controls object to the government mandating the use of key recovery, arguing that foreign companies are not bound by such restrictions. They argue that customers who do not want U.S. law enforcement or national security agencies having access to decryption keys will buy encryption products from foreign suppliers. They insist that the U.S. government cannot control the availability of strong non-key recovery encryption products, since they already can be procured from foreign suppliers, or downloaded from the Internet. They assert U.S. policies simply ensure that U.S. companies will lose market share to foreign competitors and will not achieve the overall objective of assuring law enforcement access to encrypted information of criminal groups. They point out that drug cartels, for example, could develop their own encryption products rather than buying commercially available products that would allow governments to access the keys.

Proponents of export controls concede that some criminal groups may develop their own encryption, but insist that at some point they will have to interact with mainstream companies (such as banks or airlines). If the mainstream companies are using key recovery-based systems, this would provide an opportunity for law enforcement to access at least some of the groups activities. They also point out that even though law enforcement agencies have been allowed to tap telephone lines for decades, criminals still use telephones because the infrastructure is already in place, easily used, and less costly than building an alternative system for their own use. As for foreign competition, proponents argue that although some strong non-key recovery products are available from the Internet or foreign suppliers, they are not widely used and some are not as strong as their advertisements claim.

Some cases involving encryption export controls have been the basis for legal action. One involves University of Illinois Professor Daniel Bernstein and his attempts to publish, both in print and on the Internet, the source code for his Snuffle encryption algorithm. The government argued that the export required a license under the Arms Export Control Act (AECA) and the International Traffic in Arms Regulations (ITAR) through which AECA is implemented. On April 15, 1996, U.S. District Judge Marilyn Patel ruled that computer source code is "speech" and protected under the Constitution. On December 18, she further ruled that ITAR represents an unconstitutional prior restraint on free speech. Following the December 30, 1996 shift in jurisdiction over commercial export products from the State

Department to the Commerce Department, Bernstein's lawyers asked her to review the new regulations. On August 25, 1997 she ruled that the new regulations also violate the First Amendment. On June 21, 1999, the Department of Justice filed a petition with a federal appeals court for reconsideration in the Bernstein case, which on May 6 upheld a ruling that encryption source code is scientific expression protected by the First Amendment.

An opposite ruling was made in March 1996 by Judge Charles Richey in a case involving Philip Karn. Mr. Karn was denied permission to export source code on diskette even though the source code had been published in a book and hence was in the public domain. The State Department designated the diskette as a "defense article" under AECA and denied its export. Judge Richey dismissed the complaint because the AECA does not permit judicial review of what is designated by the President as a "defense article." Mr. Karn appealed the ruling, but by the time the appeal was heard in January 1997, the export regulations had changed so the case was remanded back to DC District Court. On July 7, 1998, U.S. District Judge James Gwin ruled that an individual could not challenge encryption export restrictions on the grounds that they abridge his right to free speech on the Internet.

Domestic Use

The focus of the encryption debate has at times shifted to include potential changes to domestic use policy. Current U.S. policy allows any type of encryption to be used in or imported into the United States. Administration concerns that attempting to change this policy would be unsuccessful was a factor in its choice of using export controls to influence what encryption products are available for domestic use. FBI Director Freeh's testimony to the Senate Judiciary Committee on September 3, 1997 heralded a shift in the debate toward the possibility of requiring that key recovery be built into products manufactured in or imported into the United States, and possibly enabled by the manufacturer, not only the user. The Administration's policy, however, has not changed on this issue.

In September 1999, questions were raised over the discovery of a software element, labeled NSAKey, in the security code of the Microsoft Windows operating system. Microsoft stated that the element was a back up used for authentication of encryption components if the first key is damaged. Some question whether the key might enable the National Security Agency (NSA) to gain access to Windows operating systems. While it is unlikely that Microsoft would collaborate surreptitiously with NSA, the dispute highlights the level of tension between industry and privacy rights groups over key recovery practices.

LEGISLATION

H.R. 850 (Goodlatte)

Security and Freedom Through Encryption (SAFE) Act. Similar to H.R. 695 from 105th Congress, with changes including creating criminal penalties for the use of encryption to conceal criminal conduct, and directing the Attorney General to compile examples in which encryption has interfered with law enforcement. Introduced February 25, 1999; referred to Committees on Judiciary and on International Relations; referred to Subcommittee on Courts and Intellectual Property March 3; Committee mark-up March 24; reported by Committee (H.Rept. 106-117, Part I) April 27 without amendment; referred jointly and sequentially to Committees on Armed Services, Commerce, and Intelligence, for a period ending not later than July 2 (later extended to July 23); reported by Committees on Commerce June 23 (H.Rept.106-117, Part II); International Relations July 19 (H.Rept. 106-117, Part III); Armed Services and Intelligence July 23 (Parts IV and V). Placed on the Union Calendar July 23.

H.R. 2086 (Sensenbrenner)

Networking and Information Technology Research and Development Act. Contains provision for an NSF study comparing availability of encryption technologies in foreign countries to such technologies subject to export restrictions in the United States. Introduced June 9, 1999; reported (amended) by the Committee on Science (H.Rept. 106-472, Part I) November 16; referred to Committee on Ways and Means.

H.R. 2413 (Sensenbrenner)

Computer Security Enhancement Act. Contains provisions for establishment of a public Key Management Infrastructure and encryption standards for federal computers. Introduced July 1, 1999; referred to Committee on Science.

S. 798 (McCain)

Promote Reliable On-Line Transactions to Encourage Commerce and Trade (PROTECT) Act of 1999. Intended to promote electronic commerce by encouraging and facilitating the use of encryption in interstate commerce consistent with the protection of national security. Introduced April 14, 1999; referred to Committee on Commerce, Science, and Transportation; ordered to be reported June 23 without amendment; report filed August 5 (S.Rept. 106-142).

S. 854 (Leahy)

Electronic Rights for the 21st Century Act. Intended to protect the privacy and constitutional rights of Americans, to establish standards and procedures regarding law enforcement access to location information, decryption assistance for encrypted communications and stored electronic information, and other private information, and to affirm the rights of Americans to use and sell encryption products as a tool for protecting their online privacy. Introduced April 21, 1999; referred to Committee on the Judiciary.

Chapter 14

BROADBAND INTERNET ACCESS: BACKGROUND AND ISSUES

Angele A. Gilroy and Lennard G. Kruger

SUMMARY

Broadband or high-speed Internet access is provided by a series of technologies that give users the ability to send and receive data at volumes and speeds far greater than current Internet access over traditional telephone lines. In addition to offering speed, broadband access provides a continuous, "always on" connection (no need to dial-up) and a "two-way" capability, that is, the ability to both receive (download) and transmit (upload) data at high speeds. Broadband access, along with, the content and services it might enable, has the potential to transform the Internet: both what it offers and how it is used. It is likely that many of the future applications that will best exploit the technological capabilities of broadband have yet to be developed.

There are multiple transmission media or technologies that can be used to provide broadband access. These include cable, an enhanced telephone service called digital subscriber line (DSL), satellite, fixed wireless, and others. While many (though not all) offices and businesses now have Internet broadband access, a remaining challenge is providing broadband over "the last mile" to consumers in their homes. Currently, a number of competing telecommunications companies are developing, deploying, and marketing specific technologies and services that provide residential broadband access.

From a public policy perspective, the goals are to ensure that broadband deployment is timely, that industry competes fairly, and that service is provided to all sectors and geographical locations of American society. The federal government — through Congress and the Federal Communications Commission (FCC)—is seeking to ensure fair competition among the players so that broadband will be available and affordable in a timely manner to all Americans who want it. While the FCC's position is not to intervene at this time, some assert that legislation is necessary to ensure fair competition and timely broadband deployment.

One proposal would ease certain legal restrictions and requirements, imposed by the Telecommunications Act of 1996, on incumbent telephone companies who provide high-speed data (broadband) access. Proponents assert that restrictions must be lifted to give incumbent local exchange companies (ILECs) the incentive to build out their broadband networks. Opponents argue that lifting restrictions would allow the ILECs to monopolize voice and data markets. An alternative approach, establishing "new tools" to ensure that markets are open to competitors, is also being considered.

Another proposal would compel cable companies to provide "open access" to competing Internet service providers. Supporters argue that open access is necessary to prevent cable companies from creating "closed networks" and stifling competition. Opponents of open access counter that healthy competition does and will exist in the form of alternate broadband technologies such as DSL and satellite.

Finally, legislation seeks to accelerate broadband deployment in rural and low income areas by providing loans, grants, or tax credits to entities deploying broadband technologies.

MOST RECENT DEVELOPMENTS

H.R. 1542 (Tauzin-Dingell) was introduced on April 24, 2001. The legislation seeks to ease certain legal restrictions and requirements on Bell operating companies and other incumbent local exchange companies (ILECs) providing broadband service. On April 25, the House Energy and Commerce Committee held a hearing on H.R. 1542; the Subcommittee on Telecommunications and the Internet held a markup on April 26 passing the measure, as amended, 19-14. The House Energy and Commerce Committee reported out, by a 32-23 vote, an amended version of H.R. 1542 on May 24. The House Judiciary Committee, by voice vote, reported out "unfavorably "an amended version of H.R. 1542. Two measures, S. 1126 and S. 1127, taking a similar approach were introduced in the Senate on June 28, 2001. Alternative measures (H.R. 1697, HR. 1698, H.R. 2120) taking a different approach have also been introduced.

BACKGROUND AND ANALYSES

Broadband or high-speed Internet access is provided by a series of technologies that give users the ability to send and receive data at volumes and speeds far greater than current Internet access over traditional telephone lines. Currently, a number of telecommunications companies are developing, installing, and marketing specific technologies and services to provide broadband access to the home. Meanwhile, the federal government — through Congress and the Federal Communications Commission (FCC) — is seeking to ensure fair competition among the players so that broadband will be available and affordable in a timely manner to all Americans who want it.

What Is Broadband and Why Is It Important?

The Internet has grown exponentially during the 1990s. According to a June 2001 Gartner Dataquest survey, 61% of U.S. households actively use the Internet. Today, the majority of residential Internet users access the Internet through the same telephone line that can be used for traditional voice communication. A personal computer equipped with a modem is used to hook into an Internet dial-up connection provided (for a fee) by an Internet service provider (ISP) of choice. The modem converts analog signals (voice) into digital signals that enable the transmission of "bits" of data.

The faster the data transmission rate, the faster one can download files or hop from Web page to Web page. The highest speed modem used with a traditional telephone line, known as a 56K modem, offers a maximum data transmission rate of about 45,000 bits per second (bps). However, as the content on the World Wide Web becomes more sophisticated, the limitations of relatively low data transmission rates (called "narrowband") such as 56K become apparent. For example, using a 56K modem connection to download a 10-minute video or a large software file can be a lengthy and frustrating exercise. By using a broadband high-speed Internet connection, with data transmission rates many times faster than a 56K modem, users can view video or download software and other data-rich files in a matter of seconds. In addition to offering speed, broadband access provides a continuous "always on" connection (no need to "dial-up") and a "two-way" capability — that is, the ability to both receive (download) and transmit (upload) data at high speeds.

Broadband access, along with the content and services it might enable, has the potential to transform the Internet — both what it offers and how it is used. For example, a two-way high speed connection could be used for interactive applications such as online classrooms, showrooms, or health clinics, where teacher and student (or customer and salesperson, doctor and patient) can see and hear each other through their computers. An "always on" connection could be used to monitor home security, home automation, or even patient health remotely through the Web. The high speed and high volume that broadband offers could also be used for bundled service where, for example, cable television, video on demand, voice, data, and other services are all offered over a single line. In truth, it is possible that many of the applications that will best exploit the technological capabilities of broadband, while also capturing the imagination of consumers, have yet to be developed.

Many (though not all) offices and businesses now have Internet broadband access. A major challenge remaining (as well as an enormous business opportunity) is providing broadband over "the last mile" to consumers in their homes. Currently, about 8% of U.S. households in the United States have broadband access. The vast majority of residential Internet users today use "narrowband" access, that is, they connect via a modem through their telephone wire. However, the changeover to residential broadband has begun, as companies have started to offer different types of broadband service in selected locations. According to J.P. Morgan, 73% of households have cable modem service available, and 45% of households have access to DSL. Combined, broadband availability is estimated to be almost 85%. However, only 12% of households with available access to broadband have chosen to subscribe.[1] No one knows exactly how many consumers will be willing to pay for broadband

---

[1] Remarks of Michael Powell, Chairman, FCC before the National Summit on Broadband Deployment, October 25, 2001, [http://www.fcc.gov/Speeches/Powell/2001/spmkp110.html]

service. Currently, the cost of residential broadband service ranges from about $50 and upward per month, plus up to several hundred dollars for installation and equipment.

Broadband Technologies

There are multiple transmission media or technologies that can be used to provide broadband access. These include cable, an enhanced telephone service called digital subscriber line (DSL), satellite technology, terrestrial (or fixed) wireless technologies, and others. Cable and DSL are currently the most widely used technologies for providing broadband access. Both require the modification of an existing physical infrastructure that is already connected to the home (i.e., cable television and telephone lines). Each technology has its respective advantages and disadvantages, and will likely compete with each other based on performance, price, quality of service, geography, user friendliness, and other factors. The following sections summarize cable, DSL, and other prospective broadband technologies.

Cable

The same cable network that currently provides television service to consumers is being modified to provide broadband access with maximum download speeds ranging from 3-10 million bits per second (Mbps), and upload speeds from 128 thousand bits per second (Kbps) to 10 Mbps. In practice, transmission speeds range from several thousand Kbps to 1.5 Mbps. Because cable networks are shared by users, access speeds can decrease during peak usage hours, when bandwidth is being shared by many customers at the same time. Network sharing has also led to security concerns and fears that hackers might be able to eavesdrop on a neighbor's Internet connection.

Digital Subscriber Line (DSL)

DSL is a modem technology that converts existing copper telephone lines into two-way high speed data conduits, Data transmission speeds via range up to 7 Mbps for downloading and 1 Mbps for uploading. Speeds can depend on the condition of the telephone wire and the distance between the home and the telephone company's central office (i.e., the building that houses telephone switching equipment). Because ADSL uses frequencies much higher than those used for voice communication, both voice and data can be sent over the same telephone line. Thus, customers can talk on their telephone while they are online, and voice service will continue even if the ADSL service goes down. Like cable broadband technology, an ADSL line is "always on" with no dial-up required. Unlike cable, however, ADSL has the advantage of being unshared between the customer and the central office. Thus, data transmission speeds will not necessarily decrease during periods of heavy local Internet use. A disadvantage relative to cable is that ADSL deployment is constrained by the distance between the home and the central office. ADSL is only available, at present, to homes within 18,000 feet (about three miles) of a central office facility. However, DSL providers are currently exploring ways to further increase deployment range,

Satellite

Internet access via satellite is available to businesses from a number of firms, and satellite delivered television is received by 11.8 million subscribers. Until recently, however (see below), only one satellite technology — Hughes Network System's DirecPC — had been available to residential customers for data transmission (i.e., Internet access). With an 18-inch satellite dish, installed at the subscriber's home and aimed at a geostationary satellite located above the equator, downloading speeds of up to 400 Kbps can be achieved. Until late 2000, uploading via satellite had not been possible; to transmit data, subscribers used a 56K modem to dial-up an Internet connection over their telephone line. However, on November 6, 2000, Starband Communications announced the first two-way Internet access satellite service for the home offering 500 Kbps downstream and 150 Kbps upstream. On December 21. 2000, Hughes announced the first shipments of its new two-way broadband satellite service, with advertised download rates of 400 Kbps and upload rates of up to 125 Kbps. On August 2. 2001, Hughes announced plans to market its broadband satellite Internet service (called DirecWay) to DirecTV subscribers. The service will cost between $60 and $70 per month, in addition to television service cost. Meanwhile, upgraded two-way high speed Internet satellite systems are expected to follow. Like cable, satellite is a shared medium, meaning that privacy may be compromised and performance speeds may vary depending upon the volume of simultaneous use. Another disadvantage of Internet-over-satellite is its susceptibility to disruption in bad weather. On the other hand, the big advantage of satellite is its universal availability. Whereas cable or DSL is not available to many Americans, satellite connections can be accessed by anyone with a satellite dish. This makes satellite Internet access a possible solution for rural or remote areas not served by other technologies.

Other Technologies

Other technologies are being used or considered for residential broadband access. Terrestrial or fixed wireless systems transmit data over the airwaves from towers or antennas. Though mostly used for businesses, fixed wireless Internet is beginning to be deployed for residential broadband service. Advantages are the flexibility and lower cost of deployment to the customer's home (as opposed to laying or upgrading cable or telephone lines). Disadvantages are line-of-sight restrictions (in some cases), the susceptibility of some technologies to adverse weather conditions, and the scarcity of available spectrum. In FY2000, the FCC began auctioning frequencies currently occupied by broadcast channels 60-69. These and other frequencies in the 700 MHz band are considered "prime" for wireless broadband applications. A number of wireless technologies, corresponding to different parts of the electromagnetic spectrum, also have potential. These include the upperbands (above 240Hz), the lowerbands (multipoint distribution service or MDS, below 3 GHz), broadband personal communications services (PCS), wireless communications service (2.3 GHz), and unlicensed spectrum.

Another broadband technology is optical fiber to the home (FTTH). Optical fiber cable, already used by businesses as high speed links for long distance voice and data traffic, has tremendous data capacity, with rates in excess of one gigabit per second (1000 Mbps). The high cost of installing optical fiber in users' homes is the major barrier to FTTH. Several telephone companies are exploring ways to provide FTTH at a reasonable cost. Some public utilities are also exploring or beginning to offer broadband access via fiber inside their existing conduits. Additionally, some companies are investigating the feasibility of transmitting data overpower lines, which are already ubiquitous in people's homes. While enormous data rates are possible through power lines, significant technical barriers remain.

Status of Broadband Deployment

Broadband technologies are currently being deployed by the private sector throughout the United States. According to a survey conducted by the Federal Communications Commission (FCC), as of December 31, 2000 there were 7.1 million high speed lines connecting homes and businesses to the Internet in the United States, a growth rate of 158% over the previous year.[2] According to the June 2001 Gartner Dataquest survey, just less than 25% of online households have high speed Internet access. More recent data are available from research and consulting firms which track broadband deployment in the telephone and cable industries. Kinetic Strategies Inc. estimates that 6.2 million households in the United States subscribed to cable modem services as of September 30, 2001. Meanwhile, according to TeleChoice Inc., 3.8 million DSL lines were in service in the United States by the end of September 2001.

Policy Issues

The deployment of broadband to the American home is being financed and implemented by the private sector. The future of broadband is full of uncertainty, as competing companies and industries try to anticipate technological advances, market conditions, consumer preferences, and even cultural and societal trends. What seems clear is that industry believes that providing broadband services to the home offers the potential of financial return worthy of significant investment and some level of risk.

From a public policy perspective, the goals are to ensure that broadband deployment is timely, that industry competes fairly, and that service is available to all sectors and geographical locations of American society. Section 706 of the Telecommunications Act of 1996 (P.L. 104-104) requires the FCC to determine whether "advanced telecommunications capability [i.e., broadband or high-speed access] is being deployed to all Americans in a reasonable and timely fashion." If this is not the case, the Act directs the FCC to "take immediate action to accelerate deployment of such capability by removing barriers to infrastructure investment and by promoting competition in the telecommunications market."

On January 28, 1999, the FCC adopted a report (FCC 99-5) pursuant to Section 706. The report concluded that "the consumer broadband market is in the early stages of development, and that, while it is too early to reach definitive conclusions, aggregate data suggests that broadband is being deployed in a reasonable and timely fashion."[3] The FCC announced that it would continue to monitor closely the deployment of broadband capability in annual reports and that, where necessary, it would "not hesitate to reduce barriers to competition and

---

[2] Federal Communications Commissions, *High-Speed Services for Internet Access: Subscribership as of December 31, 2000,* August 2001; see: [http://www.fcc.gov/Bureaus/Common_Carrier/Reports/FCC-State_Link/IAD/hspd-0801.pdf]

[3] FCC News Release, "FCC Issues Report on the Deployment of Advanced Telecommunications Capability to All Americans," January 28, 1999.

infrastructure investment to ensure that market conditions are conducive to investment, innovation, and meeting the needs of all consumers." The Commission's second Section 706 report (FCC 00-290) was released on August 21, 2000. Based on data collected from telecommunications service providers, an ongoing Federal-State Joint Conference to promote advanced broadband services, and the public, the report concluded that advanced telecommunications capability is being deployed in a reasonable and timely fashion overall, although certain groups of consumers were identified as being particularly vulnerable to not receiving service in a timely fashion. Those groups include rural, minority, low-income, and inner city consumers, as well as tribal areas and consumers in U.S. territories. The FCC acknowledges that more sophisticated data are still needed in order to portray a thoroughly accurate picture of broadband deployment. The FCC's third Section 706 report is due to be released by February 2002.

The FCC has also initiated a review to examine policies and rules that affect broadband deployment. Among those is an inquiry (CC 01-337), launched in December 2001, to examine the regulatory treatment of incumbent local exchange carriers in the provision of broadband telecommunications services. Comments are sought regarding what, if any, changes should be made in how such carriers should be treated for the provision of such services. Comments are due March 1; replies April 1. Meanwhile, the National Telecommunications and Information Administration (NTIA) at the Department of Commerce is in the process of developing the Administration's broadband policy.[4]

While the FCC's position is not to intervene at this time, some assert that legislation is necessary to ensure fair competition and timely broadband deployment. Currently, the debate centers on two specific proposals. Those are: 1) easing certain legal restrictions and requirements, imposed by the Telecommunications Act of 1996, on incumbent telephone companies that provide high-speed data (broadband) access, and 2) compelling cable companies to provide "open access" to competing Internet service providers. Each course of action is strongly advocated or opposed by competing telecommunications and/or Internet-related interests.

Easing Restrictions and Requirements on Incumbent Telephone Companies

The debate over access to broadband services has prompted policymakers to examine a range of issues to ensure that broadband will be available on a timely and equal basis to all U.S. citizens. One issue under examination is whether present laws and subsequent regulatory policies as they are applied to the ILECs (incumbent local exchange [telephone] companies such as SBC or Verizon (formerly known as Bell Atlantic)) are thwarting the deployment of such services. Two such regulations are the restrictions placed on Bell operating company provision of long distance services within their service territories, and network unbundling and resale requirements imposed on all incumbent telephone companies. In the 107th Congress, H.R. 1542 would lift, with exceptions, these restrictions and requirements for high speed data (broadband) transmission. Whether such requirements are necessary to ensure the development of competition and its subsequent consumer benefits, or are overly burdensome

---

[4] See speech by Nancy Victory, Assistant Secretary for Communications and Information, before the National Summit on Broadband Deployment, October 25, 2001.
[http://www.ntia.doc.gov/ntiahome/speeches/2001/broadband_102501.htm]

and only discourage needed investment in and deployment of broadband services, continues to be debated. Two other measures (ER. 1697 and ER. 1698) introduced in the 107th Congress, take a different approach than H.R. 1542. Both measures amend the Clayton Act in an attempt to ensure that markets are open to competition. In the Senate two measures (S. 1126 and S. 1127) dealing with broadband deregulation were introduced on June 28, 2001.

Provision of InterLATA Services

As a result of the 1984 AT&T divestiture, the Bell System service territory was broken up into service regions and assigned to regional Bell operating companies (BOCs). The geographic area in which a BOC may provide telephone services within its region was further divided into local access and transport areas, or LATAs. These LATAs total 164 and vary dramatically in size. LATAs generally contain one major metropolitan area and a BOC will have numerous LATAs within its designated service region.

Telephone traffic that crosses LATA boundaries is referred to as interLATA traffic. Restrictions contained in Section 271 of the Telecommunications Act of 1996 prohibit the BOCs from offering interLATA services within their service regions until certain conditions are met. BOCs seeking to provide such services must file an application with the FCC and the appropriate state regulatory authority that demonstrates compliance with a 14-point competitive checklist of market-opening requirements. The FCC, after consultation with the Justice Department and the relevant state regulatory commission, determines whether the BOC is in compliance and can be authorized to provide in-region interLATA services. To date two BOCs, Verizon and SBC Communications have received approval to enter the in-region interLATA market in specific markets. Verizon has received approval to offer in-region long distance service to its New York State, Connecticut, Massachusetts and Pennsylvania customers. SBC Communications has received approval to offer in-region interLATA services in Texas, Kansas. Oklahoma, Missouri, and Arkansas. The independent telephone companies, or non-BOC providers of local service, are not subject to these restrictions and may carry telephone traffic regardless of whether it crosses LATA boundaries.[5]

However, the FCC has established a procedure whereby a BOC can request a limited modification of a LATA boundary to provide broadband services, particularly in unserved or underserved areas. In a February 2000 decision, the FCC concluded that it had the authority "to approve targeted LATA boundary modifications when necessary to encourage the deployment of advanced services." The FCC established a two prong test when considering such requests. The Commission further stated that "particular attention" would be paid to the views of the state commission on whether the modification would serve the public interest and that such modifications would be "narrowly tailored."

Unbundling and Resale

Present law requires all ILECs to open up their networks to enable competitors to lease out parts of the incumbent's network. These unbundling and resale requirements, which are detailed in Section 251 of the Telecommunications Act of 1996, were enacted in an attempt to open up the local telephone network to competitors. Under these provisions ILECS are

---

[5] For a more complete discussion of LATAs and BOG entry into the long distance market see CRS Report RL3 0018. *Long Distance Telephony: Bell Operating Company Entry Into the Long-Distance Market,* by James R. Riehl.

required to grant competitors access to individual pieces, or elements, of their networks (e.g., a line or a switch) and to sell them at below retail prices.

Proponents' Views

Those supporting the lifting or modification of restrictions claim that action is needed to promote the deployment of broadband services, particularly in rural and under served areas. Present regulations contained in Sections 271 and 251 of the 1996 Telecommunications Act, they claim, are overly burdensome and discourage needed investment in broadband services. According to proponents, unbundling and resale requirements, when applied to advanced services, provide a disincentive for ILECs to upgrade their networks, while BOC interLATA data restrictions unnecessarily restrict the development of the broadband network. ILECs, they state, are the only entities likely to provide these services in low volume rural and other under served areas. Therefore, proponents claim, until these regulations are removed the development and the pace of deployment of broadband technology and services, particularly in unserved areas, will be lacking. Furthermore, supporters state, domination of the Internet backbone[6] market is emerging as a concern and entrance by ILECs (particularly the BOCs) into this market will ensure that competition will thrive with no single or small group of providers dominating. Proponents also cite the need for regulatory parity; cable companies who serve approximately 70 percent of the broadband market are not subject to these requirements. Additional concerns that the lifting of restrictions on data would remove BOC incentives to open up the local loop to gain interLATA relief for voice services are also unfounded, they state. The demand by consumers for bundled services and the large and lucrative nature of the long distance voice market will, according to proponents, provide the necessary incentives for BOCs to seek relief for interLATA voice services.

Opponents' Views

Opponents claim that the lifting of restrictions and requirements will undermine the incentives needed to ensure that the BOCs and the other ILECs will open up their networks to competition. Present restrictions, opponents claim, *were* built into the 1996 Telecommunications Act to help ensure that competition will develop in the provision of telecommunications services. Modification of these regulations, critics claim, will remove the incentives needed to open up the "monopoly" in the provision of local services. Competitive safeguards such as unbundling and resale are necessary, opponents claim, to ensure that competitors will have access to the "monopoly bottleneck" last mile to the customer. Therefore, they state the enactment of legislation to modify these provisions of the 1996 Telecommunications Act will all but stop the growth of competition in the provision of local telephone service. A major change in existing regulations, opponents claim, would not only remove the incentives needed to open up the local loop but would likely result in the financial ruin of providers attempting to offer competition to incumbent local exchange carriers. As a result, consumers will be hurt, critics claim, since the hoped-for benefits of competition such as increased consumer choice and lower rates will never emerge. Concern over the inability of regulators to distinguish between provision of voice only and data services if BOC interLATA restrictions for data services and ILEC unbundling and resale requirements for

---

[6] An Internet backbone is a very high-speed, high-capacity data conduit that local or regional networks connect to for long-distance interconnection.

advanced services are lifted was also expressed. Opponents also dismiss arguments that BOC entrance into the marketplace is needed to ensure competition. The marketplace, opponents claim, is a dynamic one but proposed deregulation would unsettle nascent competition in the market.

Open Access

Legislation introduced into the 106th Congress sought to prohibit anticompetitive contracts and anticompetitive or discriminatory behavior by broadband access transport providers. The legislation would have had the effect of requiring cable companies who provide broadband access to give "open access" (also referred to as "forced access" by its opponents) to all Internet service providers. Currently, customers using cable broadband must sign up with an ISP affiliated or owned by their cable company. If customers want to access another ISP, they must pay extra— one monthly fee to the cable company's service (which includes the cable ISP) and another to their ISP of choice. In effect, the legislation would enable cable broadband customers to subscribe to their ISP of choice without first going through their cable provider's ISP. At issue is whether cable networks should be required to share their lines with, and give equal treatment to, rival ISPs who wish to sell their services to consumers.

Arguments in Favor

Internet service providers not affiliated (or "bundled") with a cable service are perhaps the principal supporters of open access provisions. Their support of open access is driven by the concern that they could lose significant market share if cable broadband access becomes widely adopted in American homes. Some Internet content providers, long-distance providers, regional phone companies, and consumer groups have also expressed support for open access.[7] They argue that without open access, competition will be stifled and cable companies will be in a position to eventually monopolize and control broadband access to the Internet. Currently, competition is flourishing among an estimated 6,000 ISPs in the United States, with the result of falling prices and rising quality and diversity of services for consumers. Without this competition in cable broadband services, say proponents of open access, the vibrancy, dynamism, and growth of the Internet may suffer.

Open access proponents further point out that a closed cable network discriminates in service quality between the cable-owned Internet service providers, whose content is directly accessible, and independent Internet service providers, whose content is only indirectly available through the Internet. They also argue that content may be restricted by cable providers, and point to some cable companies' stated intention to restrict consumer access to any video material on the Internet longer than 10 minutes (presumably, say open access advocates, to prevent Internet delivered video from competing with cable television video programming).

Finally, an argument of fairness and "maintaining a level playing field" in broadband is often advanced by open access proponents. Given that telephone companies providing Internet access are required to allow open and equitable access to all ISPs, why, they argue,

[7] For a listing of open access supporters, see Web site of OpenNet Coalition [http://www.opennetcoalition.org]

should not the cable industry — which competes with telephone companies for Internet customers — be subject to the same requirements?

Arguments Against

The cable industry strongly opposes open access provisions, arguing that the legislation would impose unnecessary government regulation on their activities, AT&T. Time Warner Cable, and Cox Communications all testified against open access at congressional hearings in the 106th Congress. Cable providers argue that an open access mandate would inhibit their ongoing nationwide investment in broadband access. Government regulation, they argue, would create uncertainty in the market and make it more difficult to justify the huge capital investments that are necessary. Given that the goal of public policymakers is the timely availability of affordable broadband service to as many Americans as possible, an open access mandate, they assert, would slow the industry's progress toward achieving this goal.

Additionally, the cable industry representatives reject the argument that without open access, competition in the Internet access market will be stifled. They maintain that vigorous competition already exists with competing broadband access technologies (i.e., DSL, satellite). They point out that it is likely that market forces will eventually dictate that cable companies open their platform to competing ISPs without the need for government regulation.[8] With broadband deployment currently at a nascent and highly dynamic stage, they argue, it is not possible for government policymakers or regulators to predict future market or technological trends with any degree of certainty. Therefore, they assert, any kind of government intervention into the marketplace would be premature and ill-advised.

The cable companies also dispute the notion that they are creating a "closed network," and point out that cable modem users are free to access any content available on the Internet. In response to criticism regarding the 10-minute limit on video, cable spokespersons state that the costs of allowing unlimited video downloads are prohibitive at present. However, they assert, since cable Internet access will be subject to a competitive marketplace, worries about the effects of restrictive cable practices are unfounded because market forces will ultimately ensure that consumers will receive the services and content that they demand.

Finally, the cable companies advance their own argument of fairness. The cable industry has invested enormous amounts of money to build a cable broadband system (estimates range over $30 billion). A government requirement to modify their equipment to allow open access to possibly hundreds of ISPs would be technically difficult and expensive, they say. Why, they ask, should unaffiliated ISPs reap the benefits of cable industry investments?

Local Debate Moves to Federal Level

The arguments for and against open access have been heard on the local level, as cities, counties, and states have taken up the issue of whether to mandate open access requirements on local cable franchises. In June 1999, a federal judge ruled that the city of Portland, OR, had the right to require open access to the Tele-Communications Incorporated (TCI) broadband network as a condition for transferring its local cable television franchise to AT&T. AT&T appealed the ruling to the U.S. Court of Appeals for the Ninth Circuit. On June 22, 2000, the Court ruled in favor of AT&T, thereby reversing the earlier ruling. The

---

[8] Cable companies have announced access agreements with unaffiliated ISPs either voluntarily (e.g. AT&T broadband) or as part of merger approval conditions imposed by the FCC and FTC (e.g. AOL-Time Warner).

court ruled that high-speed Internet access via a cable modem is defined as a "telecommunications service," and not subject to direct regulation by local franchising authorities.

The debate thus moves to the federal level, where many interpret the Court's decision as giving the FCC authority to regulate broadband cable services as a "telecommunications service." However, the FCC also has the authority *not* to regulate if it determines that such action is unnecessary to prevent discrimination and protect consumers. To date, the FCC has chosen *not* to mandate open access, citing the infancy of cable broadband service and the current and future availability of competitive technologies such as DSL, and satellite broadband services. However, in light of the June 22 court decision, the FCC announced, on June 30, 2000, that it will conduct a formal proceeding to determine whether or not cable - Internet service should be regulated as a telecommunications service, and whether the FCC should mandate open access nationwide. On September 28, 2000, the FCC formally issued a Notice of Inquiry (NOI), which will explore whether or not the Commission should require access to cable and other high- speed systems by Internet Service Providers (ISPs).[9] Meanwhile, in the 106th Congress, legislation was introduced (H.R. 1685 and H.R. 1686) that sought to require cable companies to open their high-speed networks to competing Internet service providers. Similar legislation has not yet been introduced into the 107th Congress.

ACTIVITIES IN THE 107TH CONGRESS

In the 107th Congress, H.R. 1542 (Tauzin-Dingell) was introduced on April 24, 2001. The intent of the bill is to encourage the deployment of broadband services to rural and underserved areas by easing interLATA (local access and transport area) service restrictions imposed on the Bell operating companies (BOCs) and loosening unbundling and resale obligations imposed on ILECs. On April 25, 2001 the House Energy and Commerce Committee held a hearing on H.R. 1542. The Subcommittee on Telecommunications and the Internet held a markup on April 26 and passed the measure, as amended, by a vote of 19-14. The House Energy and Commerce Committee passed an amended version of H.R. 1542, on May 9, 2001 and reported the measure out of Committee, by a vote of 32-23, on May 24, 2001 The House Judiciary Committee was granted a limited referral and by voice vote reported out an amended H.R. 1542 "unfavorably." The measure now goes to the house Rules Committee to reconcile differences.

In the Senate two measures (S. 1126 and S. 1127) dealing with broadband deregulation were introduced on June 28, 2001. S. 1126 contains provisions to deregulate ILEC rules pertaining to collocation, interconnection, and network unbundling, but also contains a 5-year advanced services (broadband) build-out requirement. S. 1127, a more narrowly focused measure, provides for broader deregulation of ILEC broadband services, but is confined to carriers serving rural areas. Both measures were referred to the Senate Commerce, Science, and Transportation Committee.

---

[9] See: [http://www.fcc.gov/Bureaus/Miscellaneous/Notices/2000/fcc00355.pdf]

HR. 1542

H.R. 1542, as passed by the House Energy and Commerce Committee, amends provisions contained in Sections 271 (BOC entry into interLATA services) and 251(interconnection) of the 1996 Telecommunications Act (P.L. 104-104). Under present law, Section 271 prohibits the BOCs from offering interLATA services within their service regions until certain conditions are met. H.R. 1542 lifts these restrictions for the provision of data traffic; restrictions on voice traffic remain. The bill permits a BOC to offer high speed data service[10] and Internet backbone service[11] across LATAs within its service territory without having to meet Section 271 requirements.

H.R. 1542 also amends Section 251 of the 1996 Act by modifying regulations regarding unbundling (sharing) requirements and resale obligations. The bill prohibits, with exceptions, the Federal Communications Commission (FCC) or any state from requiring an ILEC to provide unbundled access to any network element for the provision of any high speed data service."[12] However, the bill preserves, with some exceptions, the FCC's line-sharing order that requires ILECs to share (unbundle) the high frequency portion of its copper loop to requesting carriers. One exception is made such that an ILEC is not required to provide unbundled access to the high frequency portion of the loop at a remote terminal. Furthermore, charges to a requesting carrier for access to the high frequency portion of the loop are permitted to be equal to the amount the ILEC imputes to its own high speed data service. The bill also prohibits the FCC and tile states from expanding an ILEC's line-sharing obligation, but permits the FCC and the states to reduce elements subject to or forbear from enforcing this requirement.

H.R. 1542 also contains provisions dealing with resale of advanced services. Under tile bill ILECs are required to offer high speed data services for resale at wholesale rates for 3 years. After the 3-year period the ILEC is still obligated to offer these services to competitors but only on a "reasonable and nondiscriminatory basis."

Section 4 also contains provisions to limit the regulation of high speed data services. With noted exceptions detailed in the savings provision or expressly referred to in the legislation neither the FCC, nor any state, "shall have the authority to regulate the rates, charges, terms, or conditions for, or entry into the provision of, any high speed data service, Internet backbone service, or Internet access service, or to regulate any network element to the extent it is used in the provision of any such service."

H.R. 1542 also contains provisions to provide Internet users with access to the Internet service provider (ISP) of their choice. Section 5 requires ILECs to: provide Internet users with the ability to subscribe to and have access to any ISP that is interconnected to the carrier's high speed data service; permit ISPs to acquire the facilities and services necessary to interconnect with the carrier's high speed data service for the provision of Internet access service; and permit equipment collocation to the extent necessary for the provision of Internet access service.

---

[10] H.R. 1542 defines high speed data services as "information at a rate that is generally not less than 384 kilobits per second in at least one direction."
[11] "Internet backbone service is defined as "any interLATA service that consists of or includes the transmission by means of an Internet backbone of any packets. and shall include related local connectivity."
[12] An exception is made for network elements described in 47 C.F.R. 51.319, as in effect on January 1, 1999.

Additional provisions, adopted during subcommittee markup, were incorporated into the substitute bill passed by the House Commerce Committee. These provisions would: clarify that the BOC's may not bundle or offer long distance voice services with high-speed data offerings, even if the voice services were offered at no charge; increase fines for violations of any enforcement measures contained in the bill to a maximum of $1 million per violation per day capped at $10 million for any single act; prohibit subsidies on high-speed data services ensuring parity with non-local exchange companies regarding subsidies;[13] and prevent the FCC from imposing fees, taxes, charges, or tariffs on Internet services.

Three additional amendments were approved during full committee markup. One amendment requires the BOC's to meet the following broadband network build-out schedule: 20 percent of the company's central offices in a state must be capable of providing high speed data services within 1 year of enactment of the legislation; 40 percent within 2 years; 70 percent within 3 years; and 100 percent within 5 years. A second amendment ensures that none of the provisions contained in the bill would abrogate or modify any existing carrier interconnection agreements. The third amendment prevents discriminatory treatment among ISPs with respect to special access. It requires ILECs to provide ISPs with special access within the same period of time it provides such access to itself or an affiliate. All three amendments were approved by voice vote.

The measure was reported Out of the House Energy and Commerce Committee, as amended, by a 32-23 vote on May 24, 2001 (H.Rept. 107-83, Part 1). The House Judiciary Committee was granted jurisdiction over the bill, until June 18, 2001, limited to considering the aspects of the bill pertaining to the "purview of the Attorney General tinder section 271 of the Communications Act" as amended. House Judiciary by voice vote reported an amended version of the measure "unfavorably." Prior to the vote on H.R. 1542 the Judiciary Committee did pass, by voice vote, an amendment to H.R. 1542 that would require Department of Justice approval before a BOC could enter the in-region interLATA data market and would clarify that telecommunications carriers are subject to the antitrust laws.

H.R. 1416 (LaFalce)

Authorizes $100 million in grants and loan guarantees from the Department of Commerce for deployment by the private sector of broadband telecommunications networks and capabilities to underserved rural areas. Introduced April 4,2001; referred to Committee on Energy and Commerce.

H.R. 1542 (Tauzin)

Amends the Communications Act of 1934 to prohibit any states or the FCC from regulating the provision of high speed data services. Lifts restrictions on interLATA data transmission by Bell operating companies while also removing unbundling and resale requirements for all incumbent telephone companies in the provision of high speed data

---

[13] It appears that further clarification may be needed regarding the specific intent of this amendment entitled "Prohibition Discriminatory Subsidies".

services. Requires incumbent local exchange companies to provide any Internet Service Provider with the right to interconnect with such carrier's high speed data service. Introduced April 24, 2001; referred to Committee on Energy and Commerce. Hearing held April25; markup held by Subcommittee on Telecommunications and the Internet on April 26; passed subcommittee, as amended, 19-14.Passed Energy and Commerce Committee, as amended, by a vote of 32-23,May 9,2001. Reported out of Commerce Committee (H.Rept. 107-83, Part 1) May 24, 2001. Referred to House Judiciary with limited jurisdiction May 24, 2001. Reported "unfavorably" as amended by House Judiciary (H.Rept. 107-83, Part 2) by voice vote, June 18. 2001.

H.R. 1693 (Hall)

Authorizes $10 million in each of fiscal years 2002 through 2004 for federal agencies participating in the Next Generation Internet program to conduct broadband demonstration projects in elementary and secondary schools. Directs the National Science Foundation to conduct a study of broadband network access in schools and libraries. Introduced May 3, 2001; referred to Committees on Science and on Education and Workforce.

H.R. 1697 (Conyers)

Amends the Clayton Act to ensure the application of the antitrust laws to local telephone monopolies; and for other purposes. Authorizes a five-year, $3 billion loan guarantee program to finance the deployment of broadband services to rural and underserved areas. Introduced May 3, 2001: referred to Committee on Judiciary and Committee on Energy and Commerce.

H.R. 1698 (Cannon)

To ensure the application of the antitrust laws to local telephone monopolies; and for other purposes. Introduced May 3, 2001; referred to Committee on Judiciary and Committee on Energy and Commerce.

H.R. 2038 (Stupak)

Gives new authority to the Rural Utilities Service in consultation with the National Telecommunications and Information Administration to make low interest loans to companies that are deploying broadband technology in rural areas. Introduced May 25, 2001; referred to Committee on Energy and Commerce and Committee on Agriculture.

H.R. 2120 (Cannon)

To ensure the application of the antitrust laws to local telephone monopolies, and for other purposes. Introduced June 12, 2001; referred to Committees on the Judiciary and on Energy & Commerce. Motion to report the measure defeated by House Judiciary, *19-15*.

HR. 2139 (Smith)

Authorizes the Secretary of Agriculture to make loans for the development of broadband services in rural areas. Introduced June 12, 2001; referred to Committee on Agriculture and Committee on Energy and Commerce.

H.R. 2401 (Mchugh)

Provides for grants, loans, research, and tax credits to promote broadband deployment in underserved rural areas. Introduced June 28, 2001; referred to Committee on Energy and Commerce, Committee on Ways and Means, and Committee on Science.

H.R. 2597 (McInnis)

Allows taxpayer deductions for purchase of broadband equipment and provides tax credits to providers of next generation broadband service to rural and urban subscribers. Introduced July 23, 2001; referred to Committee on Ways and Means.

H.R. 2669 (Moran)

Authorizes the Secretary of Agriculture to make loans and grants to improve access to telecommunications and Internet services in rural areas. Introduced July 27, 2001; referred to Committee on Agriculture and Committee on Energy and Commerce.

H.R. 2847 (Boswell)

Rural America Technology Enhancement Act of 2001. Provides: tax credits for broadband facilities development; rural area broadband support through the FCC's universal service fund; and loans from the USDA Rural Utilities Service. Introduced September 6, 2001; referred to Committees on Agriculture; Ways and Means; Energy and Commerce; and Education and the Workforce.

H.R. 3090 (Thomas, Bill)

Economic Security and Recovery Act of 2001. Section 902 (added by Senate Finance Committee) provides a 10% tax credit for "current generation" broadband service (defined as download speeds of at least 1 million bits per second) for rural and low-income areas, and a 20% tax credit for "next generation" broadband service (defined as download speeds of at least 22 million bits per second). Introduced October 11, 2001. Passed House October 24, 2001. Reported by Senate Finance Committee with an amendment in the nature of a substitute, November 9, 2001.

S. 88 (Rockefeller)

Provides tax credits for five years to companies investing in broadband equipment to serve rural and low-income areas. Provides a 10% tax credit for "current generation" broadband service (defined as download speeds of at least 1.5 million bits per second), and a 20% tax credit for "next generation" broadband service (defined as download speeds of at least 22 million bits per second). Introduced January 22, 2001; referred to Committee on Finance.

S. 150 (Kerry)

Provides tax credits for five years to companies investing in broadband equipment to serve low-income areas. Provides a 10% tax credit for broadband service delivering a minimum download speed of 1.5 million bits per second. Introduced January 23, 2001; referred to Committee on Finance.

S. 426 (Clinton)

Provides an income tax credit to holders of bonds financing the deployment of broadband technologies. Introduced March 1, 2001; referred to Committee on Finance.

S. 428 (Clinton)

Authorizes $100 million in grants and loan guarantees from the Department of Commerce for deployment by the private sector of broadband telecommunications networks and capabilities to underserved rural areas. Introduced March 1, 2001; referred to Committee on Commerce, Science, and Transportation.

S. 430 (Clinton)

Authorizes $25 million for the National Science Foundation to fund research on broadband services in rural and other remote areas. Introduced March 1, 2001; referred to Committee on Finance.

S. 966 (Dorgan)

Gives new authority to the Rural Utilities Service in consultation with the National Telecommunications and Information Administration to make low interest loans to companies that are deploying broadband technology in rural areas. Introduced May 25, 2001 referred to Committee on Commerce, Science, and Transportation.

S. 1126 (Brownback)

A bill to facilitate the deployment of broadband telecommunications services, and for other purposes. Introduced June 28, 2001; referred to Committee on Commerce, Science, and Transportation.

S. 1127 (Brownback)

A bill to stimulate the deployment of advanced telecommunications services in rural areas, and for other purposes. Introduced June 28, 2001: referred to Committee on Commerce, Science, and Transportation.

S. 1571 (Lugar)

Farm and Ranch Equity Act of 2001. Section 602 would authorize the Secretary of Agriculture to make loans and grants to entities providing broadband service to rural areas. Introduced October 18,2001; referred to Committee on Agriculture, Nutrition, and Forestry.

S. 1731 (Harkin)

Agriculture. Conservation, and Rural Enhancement Act of 2001. Title VI (Section 605) would authorize the Secretary of Agriculture to make loans and grants to entities providing broadband service to rural areas. Introduced November 27, 2001; referred to Committee on Agriculture, Nutrition, and Forestry. Committee report (S.Rept. 107-117) filed December 7, 2001.

Chapter 15

WIRELESS ENHANCED 911 (E911): ISSUES UPDATE

Linda K. Moore

SUMMARY

In 1996, the Federal Communications Commission (FCC) initiated the establishment of wireless enhanced 911 (E911). Enhanced 911 provides Automatic Number Identification (ANI) and Automatic Location Identification (ALI) functions for emergency calls to Public Safety Answering Points (PSAPs). As the deadline for implementing wireless E911 approached, many requests from the wireless carriers were filed with the FCC asking for a slower rollout of the service than required by the FCC. Issues currently before Congress concern the capacity of the wireless telecommunications industry and the PSAPs they serve to meet the technical requirements of E911 in a timely and cost-effective manner. Additionally, public safety organizations have called for an overhaul of the United States' nationwide 911 technology. A consensus has emerged for a coordinated effort to improve all aspects of 911 communications. This consensus appears to have gained momentum since the terrorist attacks of September 11, 2001. Congress may wish to consider its potential role in this effort.

BACKGROUND

In the aftermath of September 11, 2001, the present capability and future effectiveness of America's network of emergency telecommunications services is among the issues that require review by Congress. A recent hearing focused on the implementation of wireless enhanced 911 (E911).[1] Information provided by industry sources suggests that Congress might also address the broader issue of the effectiveness of all parts of the 911 network, given the current circumstances.

---

[1] House Committee on Energy and Commerce, Subcommittee on Telecommunications and the Internet, June 14, 2001; a similar hearing by the Senate Committee on Commerce, Science and Transportation, Subcommittee on Communications, scheduled for September 11, 2001, was cancelled.

The Federal Communications Commission and E911

Beginning October, 1, 2001, American wireless carriers are expected to meet Federal Communications Commission (FCC) guidelines for providing Enhanced 911 (E911). Wireless E911 must be made available to customers in areas where Public Safety Answering Points (PSAPs)[2] are ready to receive wireless E911 calls with location-finder technology. Carriers must meet standards for accuracy (ability of the technology to locate the caller within a specified number of meters); market penetration (for example, 25% of new handsets); and timeliness (for example, complying with a PSAP request within six months). Wireless carriers that do not meet FCC requirements for E911 and have not been granted a waiver may be fined by the FCC. Many requests for waivers have been filed with the FCC by the carriers.[3] In general, the waivers ask for more time to meet the requirements established by the FCC.

The FCC took the initiative to extend E911 to include wireless calls in 1996,[4] citing provisions of the Communications Act[5] "promoting safety of life and property through the use of wire and radio communication." To facilitate the effort to provide comprehensive 911 services nationwide, Congress in 1999 passed a law[6] that mandated 911 as the emergency number nationwide and made numerous provisions for its implementation. Among other provisions, the law requires the FCC to work with the states and the many other affected parties to deploy comprehensive wireless E911 service.

In 1996, the FCC plotted the course for reaching wireless E911 in two phases. For Phase I, the carriers were given a year to provide requesting PSAPs with automatic number identification (ANI) and location-finder capabilities using technology existing at the time. By 2001, for Phase II, the carriers were to have identified and implemented location-finder technologies (Automatic Location Identification, or ALI) with an accuracy significantly greater than available at the time the FCC promulgated its rules.[7] From 1997 through 2000, the FCC made several changes in its requirements that impacted the carriers ability to develop the needed technology.

In 1997, the FCC recognized the possibility of handset-based solutions for Phase II, whereas previously it had discussed only network-based solutions.[8] In late 1999, it provided criteria for deploying a handset-based solution, setting stricter standards for accuracy for handset-based technology than for network-based solutions.[9] In the same report, it altered the

---

[2] A Public Safety Answering Point (PSAP) is a specialized call center equipped to receive and relay emergency messages to the appropriate emergency service.
[3] See [http://www.FCC.gov/e911/].
[4] Federal Communications Commission (FCC), *Report and Order and Further Notice of Proposed Rulemaking,* CC Docket No. 94-102, FCC 96-264, adopted June 12, 1996, released July 26,1996.
[5] U.S.C. Title 47, Chapter 5, § 151, "Communications Act of 1934."
[6] P.L. 106-81, "Wireless Communications and Public Safety Act of 1999."
[7] Automatic Number Identification (ANI) recognizes and displays the telephone number from which the call is placed. Automatic Location Identification (ALI) provides — in the case of wireline — the address associated with the telephone number or— in the case of wireless — the approximate geographic location of the caller.
[8] FCC, "E911 Reconsideration Order," released December 1, 1997. Handset-based technology requires alterations to the handset and new network software, included in this category for regulatory purposes are solutions requiring new handsets and new network hardware — sometimes referred to as a hybrid solution. Solutions that work with the installed base of existing handsets and require investments in network hardware only are considered network-based.
[9] FCC, Third Report and Order," released October 6, 1999.

accuracy requirements for network-based solutions, changing the range from 125 meters (136.7 yards) to 100 meters (109.4 yards) for 67% of all calls.

In November 1999, the FCC ruled that cost-recovery programs to reimburse carriers did not have to be in place before a carrier could be required to comply with either Phase I or Phase II.[10] This placed greater emphasis on identifying location-finder technologies that had commercial uses for wireless carriers and their customers. The FCC subsequently affirmed that cost recovery through commercial applications was a viable solution for carriers, citing, for example, the offer of one small, high-tech start-up to provide E911 connectivity for free in return for the marketing rights to use location-finder technology for commercial services.[11]

In the "Third Report and Order" (1999) and the "Fourth Memorandum Opinion and Order" (2000), the FCC also changed deadlines and requirements for market penetration, responding to requests from carriers. For the most part, however, it rejected the carrier's requests for waivers. The FCC also dismissed[12] petitions filed by Motorola, Inc. and Nokia, Inc. stating that they could supply location-finder technology that met the new guidelines but not within the FCC timetable for market penetration.[13]

Public Safety Answering Points (PSAPs) and 911

There are approximately 5,000 primary[14] PSAPs in the United States, and an estimated 5,500 to 6,000 overall.[15] They provide emergency 911 services to 97.8% of the American population.[16] For the year ending December 31, 1999, there were 190 million 911 calls. For that year, 26.5% were made on wireless phones; the increasing popularity of mobile phones has since increased the number of wireless 911 calls to 30% or more.[17]

Better technology can improve efficiency in responding to wireless 911 calls and location-finder technology can save lives. A critical factor in implementing wireless E911 is the ability of the PSAPs to upgrade their systems to accommodate and utilize the location information coming from the carrier.

It appears that, presently, there are no hard statistics on the number of PSAPs that have requested Phase 11 support for wireless E911. The FCC does not require the carriers to report on PSAP requests received and there is no existing industry-wide or nationwide effort to track PSAP readiness or, more important, the population and area served by that PSAP. The Association of Public Safety Communications Officials-International, Inc. (APCO), reports that 29 of its "model communities" have requested Phase II; this number includes statewide

---

[10] FCC, "Second Memorandum Opinion and Order," released December 8, 1999. This action was upheld in a decision handed down by the US Appeals Court, Washington, D.C., June 29, 2001.

[11] FCC, "Fourth Memorandum Opinion and Order," released September 8, 2000.

[12] "Fourth Memorandum Opinion and Order," *op.cit.*

[13] Petition for Reconsideration of Nokia, inc. and Motorola, Inc.," filed with the FCC on December 6, 1999 and "Reply to Comments to Petition for Reconsideration of Nokia, Inc. and Motorola, inc.," filed March 3, 2000.

[14] A primary Public Safety Answering Point (PSAP) is typically a regional center that is equipped to work with smaller PSAPs and local emergency response services.

[15] Telephone interview with Mr. Woody Glover, Director of 911 and Communications Center Operation Programs, Association of Public Safety Communications Officials-International, inc. (APCO), Daytona Beach, FL, conducted by Linda Moore on September 5, 2001.

[16] National Emergency Number Association (NENA), "Report Card to the Nation: The Effectiveness, Accessibility and Future of America's 9-1-1 Service." Columbus, Ohio, 2001.

[17] National 9-1-1 Day Press Conference, Senate Hart office building, September 11, 2001.

PSAPs in California, New Jersey, and Montana. There are unconfirmed estimates that 500-600 PSAPs, about 10% of all PSAPs, have sent requests to local carriers. Since PSAPs are typically served by more than one carrier, however, it is possible that the 500+ number refers to the number of requests, not the number of PSAPs ready to implement phase II.[18]

911 Technology

An important feature of today's 911 service is automatic number identification (ANI). ANI technology, however, works best with digital technology. While most of the 911 selective routing centers use digital switches many PSAPs still use analog technology at their bases. According to the National Emergency Number Association (NENA), the existing wireline 911 network infrastructure is taxed to capacity. Increased digitization and improved database management systems are among the improvements needed at the PSAP level to reduce ANI failures, increase timeliness in response and call set-up, and allow for greater information capture, such as that generated by emergency signaling technology in automobiles.[19]

DIFFICULTIES IN IMPLEMENTATION

In presenting its original blueprint for implementing E911, the FCC noted that introducing the service nationwide would require coordination and "cooperative efforts by state and local governments, PSAP administrators, wireless carriers and equipment manufacturers."

From the technology reports filed by the wireless carriers at the end of 2000 and their requests for waivers throughout 2001,it appears that the desired coordination of wireless carriers and equipment manufacturers has not occurred. The FCC, as a matter of policy, did not impose technical standards for digital wireless when it auctioned spectrum for this technology in the 1990's. Consequently, several technologies have competed to become the *defacto* standard for digital wireless technology in American markets. The European Union, in the same time frame, adopted a single standard for digital wireless. Industry opinions about how to facilitate growth in wireless telecommunications worldwide are roughly evenly divided among three major viewpoints: standardization fosters growth;[20] competition fosters growth;[21] many factors, including government intervention, foster growth.[22] The model chosen by the United States for developing wireless technology was competition. In setting

---

[18] Interview with Mr. Glover, APCO, previously cited.

[19] NENA, *op. cit.*

[20] For example, Jeffrey L. Funk, Kobe University, Japan and David T. Methe, Sophia University, Tokyo, "Market- and committee-based mechanisms in the creation and diffusion of global industry standards: the case of mobile communication," in Research Policy 30 (2001) 589-610.

[21] For example, Organization for Economic Co-operation and Development (QECD), Directorate for Science, Industry and Technology, Dr. Sam Paltridge, "Cellular Mobile Pricing Structures and Trends, Paris, 2000 [http://www.oecdwash.org].

[22] For example, Harald Gruber, European Investment Bank, and Frank Verboven, University of Leuven, Belgium, "The evolution of markets under entry and standards regulation - the case of global mobile telecommunications," working paper, revised 2001.

timetables and performance requirements for wireless E911, the FCC apparently did not factor into its calculations the difficulties of developing ALI solutions in a market characterized by multiple, competing standards.

From information supplied by APCO and NENA, noted above, it appears that the necessary level of readiness for PSAPs has not been achieved. NENA has asked the FCC to begin to track PSAP Phase II requests,[23] something that has not been done to date. It is not clear from the language of existing FCC rules whether all elements of Phase II must be in place at the time of the request or whether the request can be made in anticipation of readiness within the stipulated time frame. The FCC's Wireless Telecommunications Bureau (WTB) has sought comment on the timing of the request from PSAPs to carriers for Phase II E911 service.[24] The FCC has not yet ruled on this issue.

The PSAPs are public entities, funded at the state or local level. While the FCC did require that a funding mechanism be in place for the PSAPs before upgrading could occur, it did not mandate funding action by state or local governments. In a letter in May 2001, the WTB addressed some of the issues regarding which parties in an E911 deployment should pay for modifications to what links. The WTB emphasized that it favors "negotiations between the parties as the most efficacious and efficient means for resolving disputes for cost allocations." The letter goes on to state that differences in state laws and regulations governing 911, and the variety of existing agreements between telecommunications carriers and PSAPS, "argue against a uniform federal mandate.[25]

ISSUES

Looming large behind the issue of delays in implementing wireless E911 in a timely and cost-effective manner is the broader issue of a coherent 911 technology policy. NENA has asked Congress to consider: "ways to encourage accelerated upgrading of established 911 networks in order to provide better service to the American people;" the status of the 911 network; and "how pending telecommunications and public safety legislation may further impact 911 service." Other policy considerations might include:[26]

The extension of ALI service to include not only wireless telephones but also other wireless devises such as Personal Data Assistants (PDA's) and other hand-held devices — Palm Pilots, Visors, BlackBerries, and MP3 players;

1. The upgrading of the 911 wireline and wireless backbone and data collection points (at PSAPS or elsewhere in the system), including conversion from analog to digital systems and improved database management; the latter is of critical importance to maintain wireline location identification in a local number portability environment;

---

[23] National 9-1-1 Day Press Conference, Senate Hart office building, September 11,2001.

[24] Public Notice DA 01-1623, July 10, 2001.

[25] Sent May 7, 2001, to the E-911 Program Manager of King County, Washington. This letter can be accessed at [http://www.fcc.gov/e911/[, under "Recent FCC 911 Actions King County."

[26] Sources for these policy suggestions include the previously cited NENA report and National 911 Day Press Conference; newspaper reports about the aftermath of the September 11, 2001 attacks; comments by Eli M. Noam, Director, Columbia Institute for Tele-Information and Professor, Graduate School of Business, Columbia University, New York, NY and Roy A. Maxion, director of the dependable-systems laboratory at Carnegie-Mellon University, Pittsburgh, PA.

2. Using reports of PSAP readiness to the FCC to coordinate timely roll-outs of E911 where carriers are capable of supporting the technology;

3. Monitoring and encouraging the development of new wireless technologies and the phasing out of analog technology; newer technologies have better data capacity and provide longer battery life, both essential to wireless performance in an emergency;

4. Better management of spectrum, including policies that assure that "peak load" demand can be met in times of emergency;

5. Better contingency plans and the coordination of wireline and wireless networks to provide additional redundancy.

CONCLUSION

The telecommunications industry and the PSAPs appear to be moving toward development of a wireless E911 system that will meet the intent of Congress and the FCC but not the FCC deadlines. Issues currently before Congress concern the capacity of the carriers and the PSAPs they serve to meet the technical requirements of wireless E911 in a timely and cost-effective manner. Congress may wish to widen the scope of its inquiries to include all aspects of 911. In light of the September 11 terrorist attacks, it has been suggested that a 911 readiness campaign similar to that undertaken for Y2K be mounted, involving representatives from the public and private sectors at a level not within the scope of the FCC. These representatives might include, with the FCC, the National Highway Transportation Safety Association, representatives of the communications carriers and industry suppliers, experts in infrastructure, and the associations and officials concerned with public safety.

INDEX